AF522425

PLANT PATHOLOGY

PLANT PATHOLOGY

Amar Tyagi

ANMOL PUBLICATIONS PVT. LTD.

NEW DELHI - 110 002 (INDIA)

ANMOL PUBLICATIONS PVT. LTD.
H.O.: 4374/4B, Ansari Road, Darya Ganj,
New Delhi-110 002 (India)
Ph.: 23261597, 23278000

B.O.: No. 1015, Ist Main Road, BSK IIIrd Stage
IIIrd Phase, IIIrd Block,
Bangalore - 560 085 (India)
Visit us at: www.anmolpublications.com

Plant Pathology

First Published, 2006

ISBN 81-261-2656-6

PRINTED IN INDIA

Published by J.L. Kumar for Anmol Publications Pvt. Ltd., New Delhi - 110 002 and Printed at Mehra Offset Press, Delhi.

CONTENTS

PREFACE

Plant Pathology or Phytopathology is a very important, economically and scientifically, a most useful branch of biological sciences. It has a long history. Since early ages man has tried to grow food and improve the quality of food producting plants. Plant play an important role in our life. Man's life is always influence by plant around him. So Phytopathology gains an important part in science. The aim of this book is to provide a general overview of the field in most effective and positive manner. It is designed to meet the needs of all students. Chapters are largely independent so that the teacher may choose the desired topic. The level of presentation will be comfortable to teacher as well as student.

Various topics have been described in semi-technical language. The description have been supported with several illustrations. The subject matter and many illustrations have been taken from many books and journals. So, author does not claim any originality of the work. Author had tried hard to be accurate and up to date and realizes the impossibilities of completely avoiding error.

Author

1

Introduction

GENERAL NOTION AND NATURE OF PLANT PATHOLOGY

DEFINITION

Phytopathology or in layman term diseases of plant is derived from 3 Greek words a) Phyton–plants b) Pathos-suffering & c) logos–knowledge. So phytopathology is the study of diseases in plant and all activities in different fields concerned with the understanding of these diseases.

So, the study of nature, development and control of plant diseases is called phytopathology.

The word disease, is difficult to define in a few words. A simple dictionary meaning of disease is, "any departure from health, presenting marked symptoms, malady, illness, disorder". Several definitions of diseases in plants have been proposed. Some examples are, "a series of harmful physiological processes caused by continuous irritation of the plant by a primary agent"; "a harmful deviation from the normal functioning of physiological processes"; "a continuous, impairment of metabolism". Stakman and Harrar defined plant disease as, "a physiological disorder or structural abnormality that is harmful to the plant or to any of its part or products that reduces the economic value".

In this chapter and book we will be using many terms used in phytopathology. So, it becomes essential to discuss and define the meanings of these terms which will be included in first chapter.

Basic classification of plant diseases

Plant diseases can be classified into two heads.

(a) Simple or simple interest disease and (b) Complex or complex interest disease.

Simple Diseases

This means a disease, the propagation of which is mathematically equal to plain interest in money. There is only one generation of disease in the course. In simple type of diseases, plants and simple methods can be used for eradication of disease, for example, simple sanitation and eradication of infectious material can stop the disease. In the increase of simple interest diseases, the log [I/(l-x)] is plotted against time (where x = the proportion of diseased

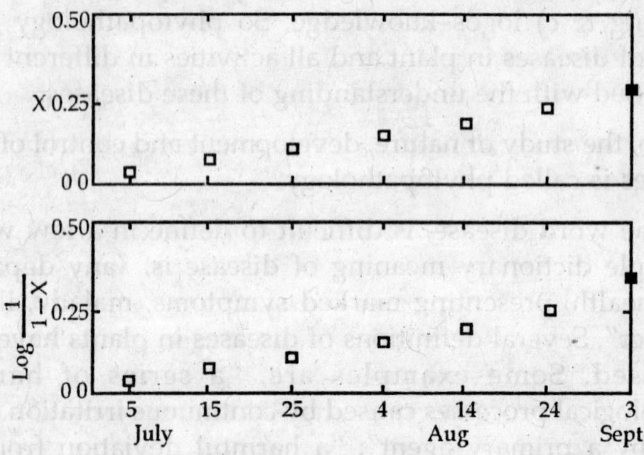

Figure. The progress of an epidemic of wilt cotton caused by *Fusarium oxysporum* f. *vasinfectum*. In the upper graph, x is plotted against time; the lower part plots Log_e [Jf(J -x)] against time. (Redrawn from J.E. Van der Plank. *Plant Diseases-Epidemics and 'Control*, Academic Press, New York and London, 1963, p.34.)

tissue), thus allowing for the fact that with increasing infection, the number or plants susceptible to infection decreases (Figure). Examples, Soil-borne diseases, such as vascular wilts and soil-borne smuts which infect seedlings and subsequently sporulate in the inflorescences of the mature plants are example of this diseases.

Complex Diseases

It means a disease whose propogation is like complex interest in mathematics. There are many generations of the pathogen in the life of the crop, In plotting the amount of disease against time, the transformation log *(x/l-x)* is used, where *x* represents the proportion of infected susceptible tissue or of, infected plants if the pathogen develops in a systematic manner (Figure). Example, Late blight of potato, stem rust of wheat.

In compound type, the epidemic spreads rapidly and lasts long timely application of fungicides should be done. The main objective should shift from sanitation to reducing the infection rate in proportion to the increase in the rate of spread and duration of the epidemic.

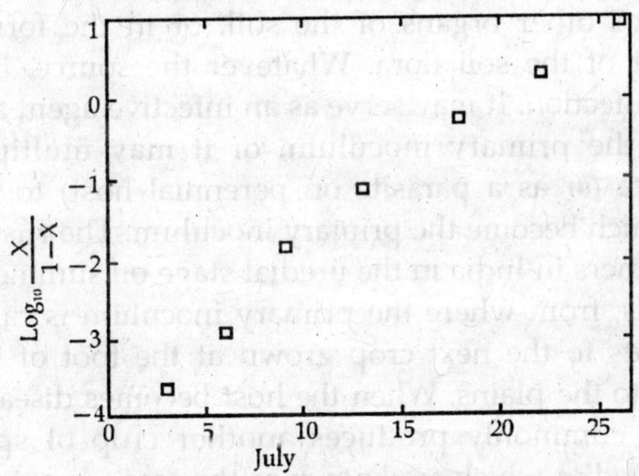

Figure. Progress of stem rust of wheat caused by *puccinia graminis tritici*. (Redrawn from J.E. Van der Plank. *Plant Diseases – Epidemics and control. Academic Press, New York and London, 1963, p.32*)

DEFINITIONS OF COMMONLY USED TERMS

Host

An organism that harbours or supports the activities of

a parasite is known as the host. Whetzel termed the diseased plant as suspect, but suspect may actually mean the plant that is susceptible or prone to disease.

Inoculum (Infectious period)

Inoculum is the infectious material that can cause disease, and it is that portion of individual pathogens that is brought into contact with the host.

Inoculum Potential (Degree of infectivity)

The term inoculum potential describes the degree or severely of infectivity. Garrett defined inoculum potential as, "the energy of a fungal parasite available *for* infection of a host at the surface of the host organ to be affected"

Primary Inoculum

The overwintering or oversummering stage may be in or on the seed of the host, in perennial wild host, in wreckage of infected plant of the previous season, in dormant spores, sclerotia or other organs of the soil, or in the form of an inhabitant of the soil flora. Whatever the source, it causes primary infection. It may serve as an infective agent and thus becomes the primary inoculum, or it may multiply as a saprophyte *(or* as a parasite on perennial host) to produce spores which become the primary inoculum. The rust fungus over summers in India in the uredial stage on summer wheat in the hills, from where the primary inoculum is carried by air currents to the next crop grown at the foot of the hills and then to the plains. When the host becomes diseased, the organism commonly produces another crop of spores or infective bodies which are known as the secondary inoculum and serve to cause are secondary infection. This last cycle may be repeated several times. The uredial pustules formed in the wheat crop as a result of primary infection become a source of secondary infection to other plants as the urediospores are blown down to these leaves and plants.

Pathogenesis

It is the succession of progress in diseases advancement

from the initial contact between a pathogen and its host to the completion of the syndrome.

Pathogenicity

It is the accomplishment or characteristic of a pathogen of being able to cause disease.

Pathotype

This is a subdivision of a species distinctioned by the common characters of pathogenicity, particularly in relation to the range of hosts.

Serotype

A serotype is a group of a pathogen in which all dividuals possess a given character of serology in common. On the basic of serological tests, differences between apparently similar organisms or viruses may be found and the subdivisions thus formed are called serotypes or strains. Such as, only one serotype was designated amongst 45 isolates of *Xanthomonas oryzae (X campestris)* on the-basis of agglutination reactions and gel diffusion tests

Physiotype

It is a population of pathogen in which all individuals have a particular character of physiology (but not pathogenicity) in common.

Penetration

The first step in the close union of the inoculum with the host. Penetration may take place in one of a number of ways which are characteristic of the organism and are sometimes influenced by the environment. The various methods of penetration are described in coming chapters. There is a clear distinction between penetration and infection. Penetration refers to the initial invasion of the host by an organism. We may have many cases of penetration without infection, *for* example; in the case of apple scab fungus *(Venturia inaequalis)* conidia, the infection hyphae penetrate readily whether or not the host variety and the fungal strain are compatible (Nusbaum and Keitt; 1938), but in the resistant host variety, the fungus is not able to develop further and cause infection due to the influence of the vital activities of the cell.

Infection

It implies the settlement of the pathogen inside the host following penetration in which a parasitic relationship between the two organisms is established.

Incubation Period (Hatching period)

This gap of time between the penetration of the host by a pathogen and the first appearance of symptoms on the host is known as the hatching period. In the case of black stem rust of wheat, the pathogen *Puccinia graminis tritici* enters the host (wheat) a day or so after the urediospores have germinated, but the actual symptoms of the disease in the form of rust pustules develop only after a week or so at 20°C.

In some cases, the hatching period can be several months.

Epidemiology

Epidemiology is a study of the factors affecting the eruption of an infectious disease.

Symptoms

Disease advancement is a long process which involves the production of secondary inoculum and progression of external or internal evidences of the disturbance of morphological and physiological processes of the host plant. These external or internal reactions or alterations as a result to a disease are referred to as symptoms.

Disease Advancement

It refers to a order of that which happens between the time of infection and complete advancement of the disease.

Syndrome (Symptoms)

Diseased plants show many valuable symptoms by

which a disease is identify. These symptoms are collectively known as syndrome.

Disease Cycle

A series of stages of development of the pathogen and the effect of the disease on the host, is called the disease cycle. When a pathogen is involved, the disease cycle is intimately associated with the organism. It is definitely distinct from the life cycle of the organism.

Infectious Organism

An infectious organism is one which may be transmitted from a diseased host plant to a healthy plant and which is capable of inciting the disease in a favourable environment.

Infectious Disease

An infectious disease is one in which a transmissible happening by chance organism is a part of causal complex, or in other words the disease can spread from a diseased to a healthy plant.

IMPORTANCE OF PLANT PATHOLOGY

People directly dependent upon plants for his survival because plants are primary source of food. The role of plant pathology is important for alleviating losses to our food, fodder, fibre and other plant products. Plants are also important to man because they utilize CO_2 in photosynthesis and release O_2.

PHYTOPATHOLOGY AND ITS EFFECT ON HUMAN LIFE

Most important example of effect of plant disease on human history is the late blight of potatoes, caused by the *fungus Phytophthora infestans.* In the period between 1840 and 1847 it resulted in a famine in Ireland where people had come to depend on potato as their major source of food. In 1845 the potato blight devastated millions of acres in Europe, the USA and Canada. So sudden and so complete was the catastrophe that in only a few days, fields which had promised abundant harvests, were transformed into blackened waste of vegetation.

As a result of this total failure of potato crop in Irland there was a great famine and about a million of people died, and almost the same number of people migrated to other continents.

The eating habits of people are very difficult to change and plant diseases appear to be responsible for many deeply implanted eating habits. While in northern and southern Europe, wheat bread is more common, in central Europe, people eat rye bread. In the southern parts of the USA people eat corn bread, while in the north wheat bread is more common. Horsfall has suggested that this situation resulted due to the ravages of wheat rust. In France between 1878 and 1882, the wine industry was threatened due to the introduction of downy mildew *(Plasnopara viticola)* from the USA. In India, the 1942 Bengal famine was perhaps largely due to the *Helmint lhosporium* disease of rice, caused by *Helminthosporium oryzae* Breda de Haan. About two million people died of starvation. According to Padmanabhan, nothing as devastating as the Bengal epiphytotic of 1942 has been recorded in plant pathological literature. In 1946-47, the wheat rust epidemic II was responsible for food shortage as well as seeds in Madhya Pradesh. The bacterial blight of rice, thought to be confined to Japan was reported. In 1951 in Bombay (Srinivasan *et al.*, 1959) and in a short time became very serious and widely distributed, attacking the commonly cultivated varieties. The disease broke put in an epidemic form in the Shahbad district of Bihar in 1963. With the Introduction of Talchung Native, the disease appeared in a severe form in 1966 throughout the country, wherever the crop was grown. Thus, with the introduction of one susceptible variety, a disease which was unknown in all but two states earlier became pandemic within two years. Coffee rust (Hemileia l astatrix) was recently discovered in Brazil where the pathogen has spread rapidly over the entire area. Coffee rust epiphytotic caused colossal losses in Brazil during 1974-1975. Thus, the disease has crossed the Atlantic Ocean to Brazil, and Columbia when it is spreading.

Chestnut blight caused by the fungus Endothia parusitica

nearly wiped out American chestnut. It is believed that the causal organism was introduced from eastern Asia at the beginning of this century. The first report of the disease came from the New York Zoological Park in 1904. A year later, the disease assumed serious proportions and the Forestry Division of the Department of Agriculture became alarmed. Although the causative agent of the disease was soon discovered, no suitable control measures could be found to control the disease. The result was that about 30 billion board feet of wood, used for furniture, flooring, poles, railroad tiers (sleepers) and coffins, was lost. The tannin industry, which was dependent on this tree, collapsed and this forced the leather makers of the USA to obtain this item from other sources. The loss of one species ,treated far reaching effects that have not been completely overcome till now. The dramatic biological and human events of this epiphytotic and the attempts to stop it have been documented recently in an excellent paper by Hepting (1974).

Another instance of a seriouslass by a fungus disease is due to the red rot of 'sugarcane *(Physalospora, tucumanensis)* which reached its peak in 1938-39 in the white sugar belt of northern India-Bihar, Punjab and Uttar Pradesh. In the badly affected areas, most of the mills could crush only 33 per cent of their normal quantity (Woodhead *et all.*, 1945). The farmers suffered tremendously.

A large number of plant diseases and the destruction they caused prove that diseases can entirely change the course of history and the economy of a country, and they have been and still are a limiting factor in crop production. The advances made in food production in developing countries due to the green revolution can be lost if proper attention is not given to plant diseases and other pests. The higher and increased yields of green revolution are based on practices which are generally favourable to attacks by plant pathogens. Today, there is an increasing awareness on the part of the scientific agricultural community that only through a deeper and more fundamental understanding of all the interacting components of agriculture, can we expect to improve our capabilities of feeding the expanding world population. Therefore, the role of plant pathologists is assuming greater importance in the presentday world. The role of plant pathology in the scientific and social development of the world has been discussed by Stakman (1958) who has also dissertated on the opportunities and social obligations in plant pathology (1964). The future and the sociology of plant pathology have been discussed by Walker (1963) and Horsfall and Cowling (1977) respectively. The scope and contributions of plant pathology have been dealt with by Houten (1959).

CAUSES OF DISEASES

Thus, it is not correct to say that micro-organism is the cause of a given disease, since it is not the sole cause.

A parasite is not synonymous with the pathogen. Many parasitics do not become pathogenic if the parasite compensates for the damage done by its presence. For example The mycorrhizal fungus is certainly parasitic on the roots of trees but it is not pathogenic. (*Rhizobium leguminosarum*).

Thrower defined a parasite as, "an organism or virus existing in an intimate association with another living organism from which it derives an essential part of the materials for its existence". In a literal sense, a pathogen is any agent which causes damage. According to *A guide to the use of terms in plant pathology* prepared by the Federation of British Plant Pathologists. A pathogen is an organism or virus capable of causing disease in a particular host or range of hosts, Most, but not all pathogens, are also parasites in that they derive the materials they need for existence from a living plant as distinct from saprophytes which derive these materials from dead organic matter.

SEVERAL LEVELS OF PARASITISM

Symbiosis

Two living organism may live in company with each other, each may be parasitic upon the other. However, they are essential to each other's development. This phenomenon is known as symbiosis. Lichens are, classical examples of symbiosis in which a fungus and an alga live in close association. The alga, through its photosynthetic activity, converts the kinetic energy of the sun's rays into the potential energy of carbohydrates which is useful for its own development and essential to the life of the fungus. The latter, through its property of producing hydrolytic enzymes, breaks down material in the substrate, and thus makes, energy available and provides minerals and organic nutrients essential to itself and the alga. Another case of symbiosis is that of the *Rhizobium* species forming root nodules in leguminous plants.

Obligate Parasites

There are several categories of parasitism. Parasites which are confined to living tissues (also known as biotrophs are called obligate parasites. Obligate is also used to denote those parasites which have not been grown in axenic cultures, that are, free from any other organism.

Saprophytes

Saprophytes are those organisms which have no relation with the living cell and secure their nutrients from dead organic tissues or from available inorganic materials.

FORMS OF PARASITISM

Following are the forms of paracitism

a) Biotrophs

These are organisms which, heedless of the ease with which they can be cultured in nature, obtain their food from the living tissues on which they complete their life cycles. Example of such rusts, smuts and mildews.

b) Hemibiotrophs

The organisms Which attack living tissues in the same way as biotrophs but continue to develop and sporulate after the tissue is dead are called Hemibiotrophs. Typical example of these are leaf-spotting fungi.

c) Perthotrophs

These are organisms which kill host tissues in advance of penetration and then live saprophytically. Sclerotium *rolfsii* is a typical fungus includes in this category.

Virulence

Aggressiveness and virulence are ambignous terms. Virulence is different from pathogenic. A pathogenic organism may cause a disease but a virulent organism is considered to have a high capacity to do so. High aggressiveness can be join together with low pathogenicity, as in the case of some obligate parasites which invade plants efficiently but cause minimum damage to it, at least in the early stages of attack.

Predisposition

It may be defined as the effect of one or more environmental factors which makes a plant vulnerable to attack by pathogen. It is a process which antedates penetration and infection.

Susceptibility

It is the incapacity of a plant to repel the effect of a pathogen or any other damaging factor. Susceptibility or a given individual may be increased or decreased by environmental factors. It may also vary between individuals, varieties or species because of differences in the inherited characteristic which affect susceptibility.

Resistance

It can be described as the power and potential of a plant to resist, restrict or prevent it self from potential pathogen. It is opposite of susceptibility and may also be influenced by environmental factors. It is the ability of the plant to withstand, oppose or overcome the attack of a pathogen completely or in some degree.

Hypersensitivity

This, outcome in a violent local reaction to attack by a pathogen and the prompt death of tissue around the points of entry prevents further spread of infection. Hypersensitivity, thus confers high resistance and is sometimes used in the sense of immunity.

Symptomless Carriers

Symptomless carriers are those plants in which the agent producing disease is present but shows no symptoms. This is a case of tolerance in some plants.

Active Resistance

Active (or dynamic) resistance is "due to reactions urged by the attack", while passive (or static) resistance is "due to qualities innate in the host prior to the attack". These terms describe resistance mechanisms and do not necessarily connote any epidemiological significance.

Pathodeme

Pathodeme is that population of a host in which all individuals have a particular character of resistance in common.

Biotype

A biotype is a population of individuals which are genetically identical (literally any species subdivision which is based on principle other than morphological).

Immunity

This implies to free from control from infection by a pathogen and in this sense is an absolute quality.

Aetiology

Aetiology is the science of the cause of disease. When used in a broader sense, it includes the study of causal factors, their nature and their relations with the host.

Fungitoxic, Fungicidal, Fungistatic

Fungitoxic substances are harmful to fungi and include fungicides. The word fungicide has originated from two latin words, viz. fungus and caedo. The word caedo means 'to kill': Thus literally speaking, a fungicide is any agent which has the ability to kill a fungus. In common usage, the meaning of the word fungicide is restricted to chemicals. Hence, fungicide means a chemical capable of killing fungi. Some chemicals do not kill fungi but inhibit fungus growth temporarily. Such a chemical is called *fungistat* and the phenomenon of temporarily inhibiting growth is called *fungistasis.*

CLASSIFICATION OF PLANT PATHOLOGY

Pathology may be classified in various ways on the basis of:

a. Host plants affected, such as cereals, millets, fruits, trees, vegetables, etc.

b. Parts of the plant affected, such as roots, stems, leaves, flowers, fruits, etc.

c. Symptoms produced on the host plants, such as wilts, blights, soft rots, anthracnoses, rusts, smuts, mildews, damping off, etc.

d. The mode of spread and severity of infection, such as epiphytotic, endemic, sporadic and pandemic.

e. Causal organism/factor.

PATHOLOGY CLASSIFIED ACCORDING TO MAIN CAUSAL FACTORS

WalKer (1969) has classified plant diseases into three main types:

A. Non parasitic diseases.

B. Parasitic diseases.

C. Mycoplasmal and viral diseases based on the major causal factors.

These are further subdivided on the basis of the causal

agency or organism involved. These are as follows :

A. Non-parasitic diseases

1. Low temperature, *i.e.*. freezing injury.
2. High temperature, *i.e.* heat canker of flax.
3. Unfavourable oxygen, *i.e.* black heart of potato
4. Unfavourable soil moisture, *i.e.* flooding injury.
5. Injurious atomosphere, *i.e.* apple scald or black tip of mango
6. Lightning injury, *i.e.* damage to potato plants due to lightning.
7. Mineral excess, *i.e.* internal bark necrosis of red delicious apple.
8. Mineral deficiency, *i.e.* red leaf of cotton or khaira disease of rice.

B. Parasitic diseases

1. Bacteria, *i.e.* bacterial blight of paddy.
2. Plasmodiophoromycetes, *i.e.* club root of crucifiers.
3. Chytridiomycetes, *i.e.* wart of potato.
4. Oomycetes, *i.e.* late blight of potato.
5. Ascomycetes, *i.e.* ergot of pearl millet.
6. Basidiomycetes, *i.e.* pink disease of citrus.
7. Deuteromycetes, *i.e.* blast of rice.
8. Phanerogams, *i.e* diseases due to giant mistletoe.
9. Animal parasites (Nematodes, *i.e.* ear cocle of wheat; insects, *i.e.* insect galls of various trees).

C. Mycoplasmal and viral diseases

1. Such as little leaf of brinjal and tobacco mosaic

CLASSIFICATION OF PLANT PATHOLOGY ON THE BASIS OF SPREAD AND SEVERITY OF INFECTION

Endemic and Epidemic diseases

A disease is termed as endemic when it is constantly

present in a moderate or severe form and is confined to a particular country or district. In plant pathology, this term is generally applied to simple interest diseases which are either indigenous or of ancient introduction.

Epidemic or Epiphytotic Diseases

An epidemic or epiphytotic disease occurs in a wide area in a destructive form. The pathogen may by present, as in endemic diseases, but the environmental factors responsible for the advancement of the disease occur only periodically, for example, cereal rusts, powdery mildews, etc.

Sporadic Diseases

Sporadic diseases, in reality, belong to the epidemic group. The term is applied to those diseases which occur at very irregular intervals, locations and in relatively fewer instances.

Pandemic Diseases

These occur allover the world (omni present) and result in mass mortality, for example, late blight of potato.

2

The History of Phytopathology

a) A study of the history of helps to make the subject clearer. We get a better perspective of the subject; we come to know the contributions in that field; the problems that were encountered and the manner in which they were tackled. We can follow the method set by previous scientists while studying.

b) Whetzel (1918) has given a useful account in his book *Outline of the History of Plant Pathology*. Large (1940) has dealt with the subject in his remarkable book, *The advance of the Fungi*. The history of plant pathology in the ages has been discussed by Oriob (1971). There are several other reviews, such as *Plant Pathology: Changing Agricultural Methods and Human Society* by TG ten Houten (1974). *Landmarks in Plant Virology: Genesis of Concepts* has been recently discussed by Markham (1977).

c) Largest number of botanists specializes in mycology and plant pathology. This is due to the fact that fungi, bacteria and actinomycetes yield antibiotics; they have a role in the decomposition of organic matter, fermentation, nitrogen fixation; they are sources of food and lastly, some of these organisms are responsible for plant diseases which cause great annual losses throughout the world. It is estimated that in India there is an annual loss of 600 crores of rupees as a result of plant diseases.

d) The earliest known writings give references to plant

diseases. Plant diseases preceded man on earth and that ever since man has practiced agriculture, they have taken their toll of crops. There are references in the Vedas (1500 B.C.) to plant diseases and the methods of control. The first book in which plant diseases have been discussed is *Vraksha Ayurveda,* written by Surapal in ancient India.

e) Theophrastus, who lived from 370 B.C. to about 286 B.C., was the first botanist to study and write about the diseases of trees, cereals and legumes. Although his approach was observational rather than experimental. He noticed that some kinds of plants are more liable to certain diseases than are others. He was not aware that diseases are caused by micro-organisms, and it was generally accepted that fungi and various other organisms observed in association with diseased or decaying plants, arose spontaneously from plants or from the environment. Comparatively little was added to the knowledge of botany or plant pathology for nearly two thousand years after the time of Theophrastus. The period between 300 A.D. and 1300 A.D. is said to be the middle or 'dark ages' in history. There were petty wars, religious dogmas and superstitions, and there was practically no increase in the knowledge of plant pathology during that period. Then came the Renaissance from the 14th to the 16th century A.D. It is during this period that printing (1440) and the microscope (1685) were invented. The compound microscope was invented by Anton van Leeuwenhoek, a linen draper, surveyor and wine gauger of Delft, Holland, and Robert Hooke, an English scientist observed the cell for the first time under the microscope in 1660.

MIDDLE PERIOD

Benedict Prevost (1755-1819) was a Swiss Professor of Philosophy at the Academy of France in 1807, he observed the germination of the spores of the wheat bunt organism and conceived the idea that this organism penetrated the young wheat plant and was the actual cause of the disease. Prevost's experiments provided the first proof and interpretation of the role of a micro-organism in the causation of disease. Prevost also demonstrated the control of smut by steeping seeds in a solution of copper sulphate. His - contemporaries did not accept his views until the devastating potato blight in 1845 which attracted the attention of mycologists and plant pathologists to plant diseases. Thus, Matheiu Tillet, Felice Fontna and Benedict Prevost were the three experimenters who were definitely ahead of their times in their thinking with regard to the causation of plant diseases.

THE MODERN PERIOD

In 1845, the devastating epidemic of the potato blight in North Europe and particularly in Ireland focussed attention on plant diseases and gave a great and stimulated interest to the Subject of plant pathology. Much of the early literature on this disease is found in the *Gardener's Chronicle*.

Scientists at that time were unable to agree on the cause of the disease or suggest a successful remady. The autogenetic theory of disease was still dominant, although opposition to this was rapidly increasing. Dr. John Lindley, Editor of the *Gardeners Chronicle* and Professor of Botany at the University College, London, was the leader of the autogenetists. *Mi* Berkeley (1803-1889) who was then the most prominent British mycologist, was at first somewhat reserved *in* supporting the parasitic theory, but advocated it strongly in 1846. Morreu in Belgium, experimented with disease in 1845. He described the disease and fungus and regarded the fungus as the major cause of the disease, although he lacked convincing evidence. That the fungus was the real cause of disease and not the result, was proven experimentally by Speerschneider in 1857 and De Bary in 1861 and 1863

After the publication of the *Origin of Species* evolution became accepted as a fact and the systems of classifications of both plants and animals were given a phylogene I tic slant.

Anton De Bary (1831-1888) was an outstanding figure in the botanical field. Much of his research was on fungi and the relation of parasitic fungi to their hosts. His mycological interests were more biological and physiological. From his investigations on life histories, para, sites and saprophytes, the nature of lichens, etc. He made new and important discoveries. Heteroecism in the Uredinales was made clear by his experiments and he gave accounts of the development and sex in a number of Phycomycetes and Ascomycetes. He was the first to indicate the nature of obligate and facultative forms. His well-known text book, *Morphologie and Physiologie der Pilze, Flechten, und Myxomyceten* in 1866 and its second edition in 1884, records the broad classification of fungi. He is truly regarded as the 'founder of modern mycology.

In his studies on the *Sclerotinia* disease of carrots and other vegetables, De Bary noted that the host cells are killed in advance of the pathogen and that the juice from the rotted tissue can break down healthy tissues of the host. On the other hand, if the juice is boiled, it has no effect on the healthy tissues. De Bary concluded that the pathogen produces enzymes that bring about the degradation and killing of plant cells. De Bary's noteworthy contributions to plant pathology were his studies on the Peronosporaceae and the diseases they incite, especially the late blight of potato, the discovery of heteroecism in the rusts and his work on *Sclerotinia sclerotiorum* (1886) which opened up an entirely new field of investigation, that is, the physiology of parasitism.

CORROBORATION OF PREVOST'S WORK

In the decade of 1840, the Tulasne brothers (Louis Rene Tulasne and Charles Tulasne) who were botanists at the Museum of Natural history in Paris, began extensive morphological studies on fungi. They made drawings of rusts, smuts and Ascomycetes which are regarded as classics in mycology. LR Tulasne is often called the 'reconstructor of mycology'. In 1847, the two brothers published studies on wheat bunt fungus and confirmed the observations of Benedict Prevost with regard to the causal organism of the bunt disease of wheat. Julius Gottheif Kuhn was a contemporary of De Bary.

He was initially a farm manager. He published his famous text book for farmers in 1858 entitled, *The Diseases of Cultivated Crops: Their Causes and Their Control.* This was the first text book to be published in which fungi were regarded as causal factors in plant disease. The diseases were grouped according to causes: unfavourable climatic and soil conditions; influence of animals; and parasitic plants. Nematodes were also included as they were known as plant parasites since the time of Needham.

HISTORY AND PRINCIPLE OF SPONTANEOUS GENERATION

(a) This theory held that the living organisms could arise spontaneously from inanimate or non-living matter. In 1748, Needhm, an English clergyman and a naturalist, supported the theory of spontaneous generation. To prove his point, he boiled meat in a flask which was corked. He demonstrated that the contents had spoiled due to micro-organisms. But while he killed the vegetative bacteria, the bacterial spores were not killed at the temperatures used. In 1775, Spallanzani, an Italian investigator, challenged Needham and proved that the decay of meat by bacteria and other lower forms of life could be prevented by heating the material in a flask which was sealed in such a way as to exclude contamination from the air. His experiments supported his contention. Others failed to get similar results consistently.

Nevertheless, the views put forward by Spallanzani at that time were not accepted. Lavosier (1743-1794) discovered the presence of oxygen in the air and its indispensability to sustain macroscopic forms of life. In 1775, he argued that this essential gas is vital for life, and therefore, sealed flasks that did not permit the entrance of oxygen were incapable of allowing spontaneous generation to occur.

(b) Schwan (1837) and Cagniard Latour (1838) discovered the yeast cell and maintained that it caused fermentation. On the other hand, in 1839, Berselius, a Swedish chemist, strongly defended the mechanistic explanation of fermentation and he was supported by Liebig, a German agricultural chemist. It was in 1860 that Pasteur furnished irrefutable evidence that micro-organisms arise from pre-existing living entities and that fermentation is a biological phenomenon rather than a purely chemical one.

(c) Pasteur began his life as a chemist, He worked as a teacher at Strassburg and then at the Ecole des Beaux Arts in Paris. He never studied medicine, though his discoveries about the protozoan nature of the silkworm and the confirmation of the bacterial origin of anthrax buried for ever the controversy over the theory of spontaneous generation.

(d) The germ theory in relation to diseases of man and animals was established in 1876 by Robert Koch. Koch was a Germen physician and a co-worker of Pasteur. The first final proof of the causal relation of the anthrax bacillus to anthrax disease was published by Robert Koch in Germany in 1876. Koch was trained as a physician; however he turned his attention to the new science of bacteriology. He began his work by establishing laboratory procedures and techniques that would yield reproducible results. Koch established the principles of pure culture technique, although it was Lister, the English surgeon who pioneered work in the field of aseptic surgery.

He was impressed by Pasteur's work and kept it as a basis of his own research. He was the first man to culture a bacterium capable of souring milk. In the medical field, Robert Koch discovered the bacterium that causes tuberculosis. He enunciated certain rules or criteria that should be satisfied before the identity of the disease-producing organism in a particular disease can be established. These rules are known as' Koch's postulates and can be briefly summarized as follows:

a. A specific organism must always be associated with a disease.

b. The organism has to be isolated in pure culture.

c. The organism must be identified.

d. The organism must produce disease in a healthy susceptible host.

e. The organism must be isolated from the experimental host, again in pure culture, and its identity established.

PLANT PATHOLOGY IN THE TWENTIETH CENTURY

Major starts at the end of 19th century gave a boost to study of plant pathology in 20th century. The rediscovery of Mendel's laws in the beginning of the century had its impact. Some of the major trends were in the following areas:

a. Genetics

b. Physiological trends in plant pathology.

c. Bacteriology.

d. Virology.

e. Ecological studies of pathogens.

f. Tissue culture in pathology.

g. Effect of environment.

h. Resistance in plants.

i. Biochemistry and physiology of the diseased plants.

j. Fungicidal research.

k. Mycoplasma group of pathogens etc.

These major trends are now discussed in brief.

PHYSIOLOGICAL TRENDS IN PLANT PATHOLOGY

The physiological trends in plant pathology can be divided into three aspects:

A. Studies on physiogenic diseases.

B. Work on relationship of the environment to infectious diseases.

C. Investigations into the physiology of parasitism.

Prevost (1807) demonstrated experimentally and interpreted clearly the indirect or secondary causal role of the environment in the etiology of parasitic disease. HM Ward emphasized the role of the environment in the epidemiology of coffee rust. He may be reckoned as the father of tropical plant pathology. He was the first to point out that the exclusive cultivation of a crop over unbroken areas is one of the most important prerequisites for an epidemic.

De Bary in his last work, pioneered research on the physiology of parasitism. De Bary worked On *Sclerotinia sclerotiorum* and the disease it causes to carrots and other plants. He demonstrated that the pathogen killed the cells of the host in advance of the invading hyphae of the pathogen. He found that juice from rotted tissue could rapidly break down healthy host tissues. This capacity to break down tissues was completely lost after boiling. Williami Brown's papers on the physiology of parasitism form the foundation of further researches in the role of enzymes in plant pathogenesis.

After the second world war, mainly on the initiative of the American, Japanese, Soviet, English and Indian researchers, the so-called physiological era commenced.

EFFECTS OF ENVERONMENT ON PLANT DISEASE

L.R. Jones pioneered the study of the effect of soil temperature on cabbage yellows at. He developed soil temperature tanks to see the effect of soil temperature on the development of disease. For this type of experimental approach, a thermostatically controlled water bath was devised into which water tight vessels containing plants growing in soil or other substrate could be inserted. This apparatus is known as the Wisconsin soil temperature tank.

Before the acceptance of the germ theory in relation to plant disease, the environment was considered to be the major causal factor. It is now well known that in cases where a parasitic micro-organism is part of the causal complex of a disease, the environment plays an important part.

GENITICS STUDY OF HOST PLANT AND PATHOGENES

In England, as early as 1905, RH Biffen began breeding cereals resistant to yellow rust. The genetics of the host plant and pathogen were studied extensively with regard to the inheritance of resistance and virulence, when Biffen's rust resistant varieties succumbed to infection in Australia, and the fact of genetically bound disease resistance was doubted. It has been demonstrated by crossing formae specialist (Stakman *et al.*, 1930) and pathotypes (Newton *et al.*, 1930) of *Puccinia graminis* that the virulence of the pathogen is also inherited, according to Mendel's laws. The same is true for various other pathogens (Person and Sidhu, 1971). The demonstration by Biffen (1905) of the Mendelian nature of the inheritance became the starting point for systematic genetic analysis of resistance. An increase in the knowledge of the sexuality of fungi by Blakeslee (1904), Craigie and Dodge (1927) enabled a genetic study of pathogenic fungi. Flor (1955) explained host parasite interaction in flax rust caused by *Melampsora lini.* He gave a gene for gene hypothesis to explain this reaction Van der Plank (1968) gave the concept of horizontal and vertical resistance in plants. Other pioneers in the field of breeding disease resistant varieties are WA Orton and LR Jones of the USA and BP Pal and KC Mehta of India.

FUNGICIDAL RESEARCH

The earliest known fungicide is sulphur. The history of the advancement of fungicides is fascinating. In the 20th century, several biochemists and organic chemists along with plant pathologists started investigations on this subject. Fungicides in the 20th century, especially in the 1930 period, were mercury or Bordeaux mixture. From the Bordeaux mixture developed the inorganic insoluble copper fungicides, such as cuprous oxides or oxychloridtes. The emphasis shifted from the oxides of copper to the organic fungicides the organo-mercurial, the dithiocarbamates and bisdithiocarbamates. WH Tisdale and Williams (1934) revolutionized the study of organic fungicides by discovering alkyldithiocarbamates. Alkylene bisdithiocarbamates were discovered by Diamond, Heuberger and Horsfall in 1943. Interest in the chemotherapy of plants has been generated and work on an interesting group of compounds, such as chloranil and systematic fungicides and antibiotics is of current interest to workers in this field. The mechanism of fungicidal action has been explored by Horsfall (1945, 1956), Sisler and Cox (1960), Sisler (1969), Rich (1960), Owen (1963) and others. In 1956 JG Horsfall published an authoritative book entitled *Principles of Fungicidal Action.* An informative book on systematic fungicides was edited by RW Marsh in 1972.

BIOCHEMISTRY AND PHYSIOLOGY OF THE DISEASED PLANT

In recent years integrated study of cause and effect has been in main stream rather than symptomato Lieal approach. It is now becoming increasingly clear that a sufficient knowledge of plant physiology and biochemistry is essential for interpreting the metabolic changes in plants during pathogenesis is a matter of fact, the biochemical and pathological trends of research started about 25 years ago and were founded on the physiological research of host, parasite relations. These modern trends of research intend to unravel the physico-biochemical processes of the diseased plant (such as photosynthesis respiration, nitrogen metabolism and growth). They wish to determine the true causes of cell-proliferation, the possibilities for the inhibition of virus multiplication, the mechanism of fungal toxicity; the mode of action of toxins; the role of hormones or hormone-like substances in pathogenesis, as well as to elucidate the virus biosynthesis. The importance of extracellular enzymes in the etiology of the disease is emphasized by workers currently interested in this area of investigation.

RESISTANCE IN PLANTS

The fact that many plants show a natural resistance to disease has long been used by plantigeneticists in their breeding programmes. Rapid advances in modern biochemistry have now made it possible for plant pathologists to investigate the biochemical basis of this resistance.

The biochemical nature of induced resistance is still not very clear. It is an interesting and important phenomenon in both virus and bacterial diseases. A significant discovery by Loebenstein showed that as a consequence of viral infection, interferon-like substances are produced in the host plant. In infections by animal viruses, similar antiviral substances appear in the host plant. An ever increasing body of research on this class of compounds has been accumulating during the last decades (Cruickshank. 1963; Kuc', 1972) but more convincing evidence as to how phytoalex ins play deciding role in host pathogen interaction, leading to resistance is still lacking.

TISSUE CULTURE IN PLANT PATHOLOGY

The biggest break in study of plant pathology is the study of tissue culture. It has enabled to grow organs, tissues and free cells alive *in vitro* for almost an indefinite period. In the early 20s a branch of conventional tissue culture-organ culture-made its appearance and has since established itself as a method of choice in many areas of plant physiology and plant pathology, biological and medical research, biochemistry of drugs, antibiotics, hormones, virus research, etc.

ECOLOGICAL STUDIES OF SOIL-BORNE FUNGAL PLANT PATHOGENS

Initially it was thought that a knowledge of the life cycle of a plant parasite was sufficient to ensure its speedy control. Pathologists have now realized that a wider outlook is required, which embraces the entire dynamics of microbial ecology, and that a knowledge of life cycles is only a prerequisite to the proper understanding of a soil-borne fungus pathogen in relation to its environment. The surroundings of an organism in the soil are biologically complex and the interactions of the organisms themselves produce balance. The most important worker in the field of root disease investigations is *Dr* SD Garrett. To paraphrase Garrett (1970), his subject has opened for him, and for others too, a window upon a world wider than that of the diseased plant. This world is that of the living soil, which is the I source of all terrestrial plant life, and therefore, of direct concern to every human being and of interest to every biologist. Inside the soil, many kinds of living organisms are interacting continuously with one another and with the root systems of green plants. Thus, diseases of the root system cannot be studied in isolation as a simple interaction between the root system and the pathogen; hence, the root-disease investigator should be concerned with everything in the soil, such as changes in the soil texture, temperature, moisture, *pH* value and other biological effects.

VIROLOGY

German born Adolph Mayer (1886) worked in Holland on a serious disease of tobacco (tobacco mosaic) and this work can be said to be the alpha of virology. He showed that the juice from infected plants, if applied to healthy Plants, could reproduce the disease. In the same way, EF Smith (1891) showed that the peach yellows was a contagious disease and could be bud-transmitted. In 1862. Dimitri Ivanowski, a Russian, showed that the casual agent of tobacco mosaic could even go through a chamberland filter that withholds bacteria. It was Beijerinck, a distinguished Dutch microbiologist (1898) and Professor of Microbiology at the Technical University of Delft, Holland, who demonstrated that it could diffuse through an agar membrane and concluded that the tobacco mosaic was caused by a noncorpuscular *contagium vivum fluidum* which he called virus. In 1935, the first major contribution was made regarding the nature of viruses when Stanley obtained a crystalline protein which he thought to be gamma globuline. It was in 1938 that Bawden and Pirie discovered the real nature of the tobacco mosaic virus. In 1956, Gierrer and Schramm (1956) proved that the nucleic acid fraction of the virus is actually the infectious agent. Within the attacked cell, virus somehow mobilizes nitrogenous and other compounds of the plant and replicates itself. The relationship between an insect and a virus was discovered by a Japanese farmer, Hashimoto, who worked in 1894 with the rice dwarf disease and the leaf hopper *Nephotettix apicalis* var. *cinticeps*. This information was however not available to the western world for quite some time (Smith, 1974). About 1907, the three American workers—Ball, Adams and Shaw-proved the connection between the curly top of sugar beet and the leaf hopper *Eutettix tenella*.

During the last decade, much more information on the relationship of plant viruses with their vectors has accumulated. It is now clear that in addition to the insects which transmit plant viruses, other organisms such as mites, nematodes, worms, and fungal spores have all been associated as vectors. Other recent advances in the study of plant viruses include and increased knowledge of incomplete viruses, and the discovery of satellite viruses (Kassanis, 1966) and 'viroids' (Diener and Raymer, 1967). Potato spindle tuber virus (Diener and. Raymer, 1967) and the rice dwarf virus (Hiura *et al*., 1966) are shown to contain double stranded RNA. The algal viruses which form a very distinct group of DNA viruses are another important and interesting field of research.

The rod-shaped tobacco mosaic virus was first reconstituted from its protein and RNA components (Fraenkel-Conrat and Williams, 1955). Since then, several spherical viruses have been reconstituted. "The tide of molecular biology has now reached the shores of plant virology", says Dr Kenneth M Smith; and more precise knowledge of the ultrastructure of viruses has been obtained by electron microscopy and X-ray diffraction techniques.' A photographic rotation technique has been developed to increase the details of high resolution electron micrographs. The mechanism of replication of the viral nucleic acid and other constituents within the cells is being looked into critically, and in the case of the tobacco mosaic virus, the work of KK Reddi (1964) stands out prominently. By modifying the purine bases of RNA, Mundr Y (1959) succeeded in inducing TMV mutations; thus, he was the first to elucidate the biochemical mechanism of mutation.

BACTERIOLOGY

Thomas J Burrill (1839-1916) of the University of illinois, USA, was the first to describe a bacterial disease of plant in 1878 and he showed that the fireblight of apples and pears is caused by the bacterium known as *Erwinia amylovora.* Several other plant diseases were described shortly afterwards. Wakker (1883) showed that the yellow disease of hyacinth is caused by a bacterium. Savastano (1887) worked on the olive knot disease and EF Smith worked on the bacterial wilt of cucurbits. Smith is regarded as the most outstanding man who used the best available methods and standards in the study of bacterial diseases of plants and in establishing the modern bacteriological trend in plant pathology. Smith's name is still remembered as he resolved the controversy with the German bacteriologist, Alfred Fischer (1897, 1899), who did not think that bacteria were the primary cause of disease in plants. The subject of bacteriology has now developed considerably and has its own techniques and specialists in the field.

MYCOPLASMA

Before 1967, only fungi, bacteria, nematodes and viruses were considered as main incitants or-plant diseases. However, as early as 1898, Nocard and Roux had discovered another agent of diseases caused in animals which was similar to viruses in size but which could be cultured on artificial media. This group was known as mycoplasma. *Mycoplasma mycoiils,* the causal agent of infectious pneumonia, had been cultured by that time. In 1967, Japanese scientists found that mycoplasma-like organisms could be responsible for most of the disease of the yellows type which were earlier thought, to be caused by viruses in plants.

Mycoplasma-like organisms found in certain leaf hopper-transmitted-diseases are morphologically similar to animal mycoplasma, The cells of mycoplasma lack a cellwall and hence are pleomorphic. They do have a cell membrane (10 mm unit membrane). They are very small in size (from 100 mm to I *p.* in diameter) and because of the lack of cellwalls, can easily pass through bacteria proof filters. They contain ribosomes and thin standard DNA and are sensitive to antibiotics of the tetracycline group.

3

Genetics

GENETICS OF PLANT AND PATHOGEN

Each host-parasite interaction is a contest for survival between two organisms. In nature, there is a state of balance and harmony and surviving plants and their parasites are capable of coexistence. The attack and defense mechanisms discussed earlier are determined genetically in the pathogen and host plant, respectively. Genetical variation produces in new races of pathogens, some of which have different pathogenic abilities and may be able to attack hitherto resistant plants. In both organisms, the more effective races and varieties tend to survive, the less effective ones tend to disappear. Effective disease develops when a virulent race meets a susceptible variety but not when the pathogen is avirulent or the plant is resistant. The genetics of host parasite interaction have been discussed by Day and Ellingboe.

To control plant diseases, it is must to know not only what a pathogenic organism looks like and what it is called, but also its behaviour, and that too not only in a single plant under one set of conditions, but on population of different kinds of plants under all conditions in which plants grow.

Mutation

Mutation can be described as an abrupt change in the genetic material of a cell, which is then forwarded in a genetic fashion to the progeny. Mutations occur spontaneously in nature in all living organisms which reproduce only sexually or only a sexually and also in those which reproduce both sexually and asexually. Mutations, apparently, are the results of unavoidable but rare 'accidents' that happens during cell division and result in irregularities in the replication or rearrangement of minute parts of the genetic material of the cells. Mutations can also be produced at a high rate artificially by exposing the organisms to either physical agents, such as ultraviolet light, X-rays, γ-rays and extreme temperatures, or to highly reactive chemicals., such as the nitrogen and sulphur mustards, epoxides, peroxides, phenols, alkaloids and alkylating agents. One highly potent mutagenic chemical, often styled as a magic bullet, is N-methyl-N nitro-N nitrosoguanidine.

The pathogenic abilities of parasites as well as their morphological and other characteristics can be altered in this way and there is evidence that these are important in the development of new strains. In a haploid fungus, such mutation should have immediate expression but this may be changed by epistatic or other effects. If the cell in which gene mutation occurs contains more than one nucleus, the expression will be modified by the other nuclei present and in such heterokaryotic fungi, pathogenicity and other characteristics will be a composite effect.

New varieties attributed to mutations have been reported in *Cladosporium fulvum* (tomato leaf mold) where 'step' mutation (each mutation overcoming the protective effect of a single gene) is thought to have occurred and similar mutations have been achieved experimentally by ultraviolet irradiation. Pathogenic variation due to mutation has also been seen in *Phytophthora infestans, Puccinia graminis, Melampsora lini* and various smuts, to mention but a few. Many of these mutations are likely to be recessive and their expression will be dependent on sexual reproduction with an accompanying recombination of pathogenicity factors. Anastomosis between hyphae and germ-tubes may be followed by an exchange of nuclei and this may also permit the expression of recessive pathogenic mutations.

This could be of big significance in Fungi Imperfecti and in those Phycomycetes, Ascomycetes and Basidiomycetes in which the perfect stage is rarely formed, as in some Erysiphaceae in which cleistothecia fail to develop under certain climatic conditions, Although the inadequate information available suggests that mutations affecting pathogenicity are fairly rare, they are doubtless an effective mechanism for variations when the enormous numbers of spores produced by many pathogens are taken into account, for example. Mutation can also bring about reduced pathogenicity or other changes disadvantageous to the pathogen. The practical implications of mutations in the aggressiveness of pathogens are difficult to assess. The sudden appearance of new races of increased pathogenicity in a pathogen which has hitherto been of only moderate pathogenicity has been ascribed to mutation and this may well be true. The recent appearance of a new, highly destructive race of *Helminthosporium maydis* (Southern leaf blight of maize) in the USA may be given as an example of this.

Mutations also provide the basic pathogenicity changes which are subsequently brought out by hybridization or similar recombination mechanisms and without mutation the latter would be of a more limited scope.

Heterokaryosis

Heterokaryosis and variability in plant pathogenic fungi has been best summarized by Parmeter. Hansen and Smith defined heterokaryosis as a term which ".... precisely describes the condition of a cell containing two or more genetically different nuclei . . . . " In a later classification of the use of the term heterokaryosis, Hansen pointed out that the term may' be used whenever such a condition exists, not only in a single cell, but also when it exists in any or all cells of an individual thallus. This clearly implies that heterokaryosis involves the occurrence of dissimilar nuclei in a single cell even if it is only one cell in a thallus.

Nelson used the term "heterokaryosis" to designate the association of more than two genetically different kinds of nuclei in a vegetative cell or spore. Considerable information on the potential advantages of heterokaryosis is available in the existing literature. It is agreed by most of the workers that heterokaryosis provides a haploid organism with many of the advantages of heterozygosity enjoyed by diploid organisms. Nuclear changes could provide additional somatic flexibility with changing environments (Davis, 1966). However, the main advantage of heterokaryosis appears to be the fact that it is a requisite step for genetic recombination brought about by interchanges of whole chromosomes or through mitotic crossing over. Heterokaryosis may thus play an important role in homothallic and imperfect fungi. This aspect has been reviewed by Caten and Jinks (1966).

Heterokaryosis can be produced in several ways:

a. Gene mutation: Though these are not easily detected at gross level since the frequency of these in nuclei is very small.

b. By fusion of vegetative cells: Anastomosis is undoubtedly important in the initiation of heterokaryosis. Anastomosis, as it is related to the establishment and development of heterokaryosis, involves fusion of hyphae, movement of one or more nuclei into one or the other of the Fuesed cells and the establishment of a compatible heterokaryotic state. Failure to establish heterokaryons may involve anyone of these steps.

c. Heterokarvosis may take place at the time of spore formation in many fungi, such as inclusion of + and - nuclei in the same ascospores of *Neurospora tetrasperma* or *Podospora anserina.*

Apart from dikaryotics in smuts, rusts, and Hymenomycetes heterokaryosis has been reported in pathogens belonging to the Fungi-Imperfecti, Ascomycetes and to a lesser extent, Phycomycetes, but not all these reports are entirely convincing.

Heterokaryosis appears to be necessary for pathogenicity in heterothallic rusts, but not in infection by basidiospores as the latter are normally uninucleate and hence necessarily homokaryotic. There is considerable proof that heterokaryosis not only conditions pathogenicity but also enables the effects of mutation on pathogenicity (often recessive) to be expressed. The important of heterokaryosis in the new pathogenic races of Phycomycetes. Ascomycetes and Fungi-

Imperfecti is not so well understood. Buxton (1956) produced nearly avirulent races of *Fusarium oxysporum f. sp. pisi* by ultraviolet radiation and showed that heterokaryons were recovered by culturing from the diseased plants. Buxton concluded that some of the enhancement in the virulence of pathogens in nature may be due to heterokaryosis between weakly virulent strains.

Nutritional deficiencies can affect pathogenicity as shown by Hrushowltz, who demonstrated that repeated culturing of *H. sativum* on media supplemented with certain amino acids led to a, progressive loss in pathogenicity in some cases. The mechanism involved is not clear but such repeated culturing may have led to the selection of deficient nuclei, thalli containing such deficient nuclei being less pathogenic. The increased pathogenicity due to heterokaryosis in the mycelium would not be inherited by the spores unless the latter contained a full complement of the parental nuclei and would not obtain with uninucleate conidia. Heterokaryosis is thus unlikely to bring about any permanent inherited change in pathogenicity which would persist after sporulation, although parasexual' recombination in which heterokaryosis is the first step, could do so. Heterokaryosis has been discussed by Davis.

Heterokaryosis is certainly a way in which avirulent strains may get virulence. One of the best documented cases is that of *Thanatephorus cucumeris*. Heterokaryosis between avirulent strains, both spontaneous and induced or between avirulent and virulent strains were all highly pathogenic, in the latter case more so than either component.

Parasexual Recombination

This process in which genetical recombination occurs in the vegetative thallus in the absence of a sexual stage has been investigated chiefly in species of *Aspergillus* and *Penicillium* in the laboratory of Pontecorvo and his co-workers

The sequence of events in a complete parasexual cycle is somewhat as follows:

a. Formation of heterokaryotic mycelium.

b. Fusion between two nuclei:

c. Fusion between like nuclei.

  (i). Fusion between unlike nuclei.

  (ii). Multiplication of diploid nuclei side by side by diploid nuclei.

d. Occasional mitotic crossing over during the multiplication of diploid nuclei.

e. Sorting out of diploid nuclei.

f. Occasional haploidization of the diploid nuclei.

g. Sorting out of the new haploid strains.

The results are similar to those achieved by meiosis, a much more precise and regular process. The frequency of these events appears to be low and there is little information about the significance of this somatic recombination in nature, specially as concerns the fungi which attack plants.

Process of Adaptation

Adaptation in fungi had been studied by Buxton and more recently by Person. The possibility that the pathogens might adapt themselves to their host plants has been debated for many years. There seems to be no hard evidence that this occurs although the evidence for adaptation to chemicals is stronger. Apparent adaptation to host plants might be due to mutation or to the selection of pre-existing strains which are better able to attack the host. Changes in the enzymic capabilities of a pathogen may be mediated by the presence of appropriate substrates (adaptive enzymes) and this might modify its pathogenicity. Many years ago Ward suggested the concept of bridging host by means of which a pathogen is able to pass from a susceptible to a hitherto resistant host; for example, in *Puccinia disperse* the pathogen becomes adaptive to resistant species of *Bromus*. After propagation in some species of grass, somewhat similar changes were reported by Salmon in powdery mildews, and more recent work indicates that passage through some host plants can sometimes affect pathogenicity. Repeated passage through a resistant plant sometimes enhances the aggressiveness of a pathogen and vice versa, as has been demonstrated in several fungi and bacteria; for example, *Phytophthora infestans* and

Xanthomonas stewarti .

Changes in the pathogenicity of *F. oxysporum* f. sp. *pisi* is brought about by exudates. Although there is increasing proof in favour of parasitic adaptation the mechanism involved remains obscure. Host selection of more aggressive biotypes from a mixture or biotypes, together with adaptation, mutation-(perhaps by some mutagens in the host) and cytoplasmic variations may be involved in different cases.

Saltation

This phenomenon also described as dissociation or sectoring refers to the appearance of morphologically different sectors in 'fungal colonies. It occurs frequently in some fungi, including some isolates of *Fusarium* and *Helminthosporium,* but infrequently in others. Some such sectors remain fairly stable when sub-cultured while others are unstable. Saltation may be influenced by the compaction and thickness of the culture medium and can affect such characteristics as colony colour and topography, rate of growth and sporulation and size and shape of the spore. Pathogenicity might also be affected and this has been studied in *Fusarium* species by several investigators including Haymaker, Wellman, Weindling and Oswald. The results have been somewhat controversial saltants tended to be less aggressive than the parent cultures, but occasional ones were more aggressive and these may be of significance where the saltation rate is high. No new pathogenic races with changed host ranges have been reported in saltants. The mechanisms involved in saltation are varied and not well understood.

Cytoplasmic differences

There is a great deal of proof for the existence in fungi of genetical determinants located in the cytoplasm outside the nucleus. Inheritance of such cytoplasmic determinants differs from that of nuclear genes. Cytoplasmic inheritance has been attributed to plasma genes. Cytoplasmic RNA (possibly some ribosomes) are genetical materials detached from the chromosomes (episomes). There is little information as to the possible significance of cytoplasmic factors in bringing about variation in the pathogenicity of plant pathogens, although Johnson has described the cytoplasmic control of pathogenicity of *Puccinia graminis* f. sp. *tritici* this is a subject which needs investigation. Cytoplasmic inheritance is discussed by Jinks and Fincham and Day.

Sexual Recombination

The origin of new physiological races through meiotic recombination is well attested in many pathogenic fungi, rusts, smuts, powdery mildews and the potato blight fungus. Hybridization is likely to be encouraged by incompatible mechanism which prevents or reduces inbreeding, and many pathogens which are heterothallic.

Hybridization can be intraspecific, interspecific or even intergeneric and the resulting hybrids may have different pathogenic abilities than the parental races. Often the hybrids are intermediate in pathogenicity between the two parental races but some may be more pathogenic than others and similar considerations apply to other inherited characteristics.

Extensive recombination of genes occurs in autoecious rusts such as *Melampsora lini* (flax rust) in which all the spore stages occur in the same plant. Interspecific hybridization is discussed by Nelson (1963).

Transformation Process

In addition to sexual recombination, there are many possibilities for the production of new variants in micro-organisms by asexual means. One of these asexual type of combinations of genetic characters is bacterial transformation. The observation of Avery *et al.* called attention to this process. They were successful in converting a non-pathogenic non-capsule-forming strain of *Diplococcus pneumoniae* into a pathogenic strain such forms' a carbohydrate capsule. This was done by culturing the non-pathogenic strain in the presence of the biologically active fraction of the capsule forming a pathogenic strain. It was later demonstrated by these workers that this transformation was brought about by the introduction of purified de-oxyribonucleic acid (DNA of the donor). Thus transformation is a specific form of recombination, involving the naked genes or DNA.

Several workers have demonstrated the production of

new strains of plant pathogenic bacteria by means of transformation. Corey and Starr (1957) were able to change the type of colonies and resistance of *Xanthomonas phaseoli* (a pathogen of beans) to streptomycin by transformation.

Transduction

One of the most recently discovered processes of recombination different from sexual recombination is transduction. This process has been found in a various of bacteria, where DNA is transferred from cell to cell through the agency of temperate viruses. Bacteria need not necessarily be destroyed by their infective viruses (bacteriophages). In certain cases after ınfection of the bacterium with a bacteriophage, the phage (temperate) may produce in the host, an immunity that save multiplication of the phage and lysis of the bacterial cell. Such an association is called the iyscgenic state. Under the influence of different chemicals and physical treatments (i.e. ultra violet irradiation) lysogenic bacteria liberate their latent phages by lysis. These are infective viruses with the ability to enter the new bacterial cells. During symbiosis the phage becomes the genetic material of the host (ii) Phage DNA comes into close contract with the host DNA during symbiosis. This makes it possible for the DNA molecule, carrying the genetic factor of the donor bacterium to be transformed to a recipient bacterial cell through the vehicle of a phage particle. Pathogenicity can also be transduced by bacteriophages. Therefore, transduction is an important mechanism for genetic recombination in bacteria (iii) This interesting mechanism of genetic exchange was discovered by Zinder and Lederberg in *Salmonella typhimurium*. Its importance was also demonstrate in a plant pathogenic beacterium (*Psedudomonas solanacearum*) by Okabe and Goto. One can suppose that transduction has role in producing new biotypes of plant pathogenic bacteria in nature. However, this remains to be proved.

(iv) Viruses also how considerable variability in several characteristics, such as infectivity, longevity, symptoms produced on the host plant, host-range and specificity to vectors. These variations are possibly due to some sort of mutation process and are inherited in that they are perpetuated during the transfer of the virus from one plant to another. The mechanism involved in these changes is not quite well understood in virus resistant plants, the relationship is, in effect, between the nucleic acids of the resistant genes of the host plant and RNA of the virus particle.

PHYSIOLOGICAL DEVELOPMENT

Many factors in host-parasite interactions present the fundamental problems of plant disease. The major factor towards the elucidation of host-parasite interactions is the genetical aspect. The rediscovery of Mendel's laws of heredity in 1900 and subsequently the fast development of a basic science of genetics gave a strong momentum to the study of plant pathology. The genetic basis of disease resistance was analysed in many cases. Although the genetics of host plants received most attention initially, advances in the knowledge of the sexuality of fungi led to the development of materials and methods for genetic studies of pathogenic fungi. There is now positive evidence that the general pattern of the laws of inheritance are the same in fungi as in other groups of plants. The fungi undergo heritable changes brought about by segregation and recombination of genes, mutations, heterokaryosis, cytoplasmic inheritance, etc. Not only this, because of their great variety in sexual and asexual reproduction and their unique suitability for experimental manipulation, micro-organisms now afford the best available material for many types of fundamental studies in genetics, biochemistry, pathology and other fields. Despite all these advances, however, knowledge of the actual physico-chemical mechanisms that determine parasitism, pathogenicity or disease resistance is still very meagre. But, in a nutshell we can say that disease resistance is a genetic character as regularly inherited as any other character. However, the genetic behaviour of resistance usually differs in different cases.

Resistance to one disease does not necessarily prove that it may be resistance to any other disease. Thus, Thatcher wheat, while resistant to stem rust, is highly susceptible to leaf rust. *(P. recondita, P. triticina)* Various kinds of genetic behaviour have been observed with respect to disease resistance, for example, simple monohybrid inheritance of resistance depending on a single gene with 3:1 Mendelian ratio, has been observed in many cases, as for instance, in the cabbages to yellows *(Fusarium conglutinans).* lettuce to downy mildews *(Bremia lactusae)* and peas to wilt *(Fusarium oxysporum* sp. *pisi).* In these cases, resistance is a dominant characteristic.

Two gene segregation appear to be involved in resistance to anthracnose of beans *(Colletotriehum lindemuthianum)* and some cereal rusts. Many genes which are more difficult to analyze are known to be involved in a large number of diseases

PHYSIOLOGICAL SPECIALIZATION IN FUNGI

Intricacy of Organisms

Like all other organisms, fungi are also complex and variable. This complexity is governed by the genetic material contained in the chromosomes. The crop plants are similarly complex, for example, there are 14,000 varieties of wheat; similarly there are innumerable varieties of apples, beans, grams, roses, corn, etc.

Actions of Genes

When fungal species parasitize a number of host varieties they often develop physiological specialization, which is the development of certain characters in the fungus which are not morphological but physiological and can be detected only when the effect is noted on the host. A special species of fungus such as the black stem rust, *Puccinia graminis,* may include a number of "physiological races" which are similar in morphological structure but may be different in their capacity to produce disease. So the difference between them is physiological and not morphological. Such races are commonly encountered in many groups of parasites, such as rusts, smuts, powdery and downy mildews, etc.

Gene action expresses itself in various morphological features and it also always governs the various physiological processes of the organisms. This action is through the production of various enzymes. The interaction between the host and parasite is actually the interaction between the enzyme systems of the two and the result is the effect produced.

The details of this phenomenon can be easily explained by taking the example of the black stem rust *Puccinia graminis*. The discovery of pathogenic specialization is fundamental to studies on pathogenic variability. Eriksson showed that *P. graminis* on cereals and grasse is composed of several pathogenically different strain which he described as specialised forms (formae speciales). These are now known as varieties. There are at least six varieties of *P. graminis* which can be recognized by the size of the spores and also by their pathogenicity on different hosts:

Each variety, for example, *P. graminis tritici* is further subdivided into pathogenic units called physiological races. These races cannot be distinguished by morphological examinations but by noting the different effects they produce on various varieties of wheat. The only method by which the nature (identity) of physiological races can be known is to grow them on a number of host varieties and note the symptoms which will be different if the isolates differ in physiological capacities. In order to have a standard method for the identification of a physiological race, some host varieties are fixed, which are called "standard differentials", and the record of each race is taken. These differentials, now used to distinguish *Puccinia graminis tritici* and *Puccinia graminis avenae*, were selected about forty years ago. Whenever the necessity arises new ones are added. For identifying the races, three varietal reaction classes are used. These are resistant (R), susceptible (S), and mesothetic (M). Further degrees are recognized within these three classes, for example, immune, practically immune, extremely resistant, moderately resistant, moderately susceptible, completely susceptible, and heterogenous mixed reaction

Types of Infection

According to the degree of spotting produced, each type of infection is indicated by one of the six given symbols:

a) Type 0 Very faint, light flecks; host therefore considered immune.

b) Type 1 Uredosori very minute; host strongly resistant.

c) Type 2 Small to medium; host resistant.

d) Type 3 Medium size; host moderately susceptible.

e) Type 4 Large and confluent; host very susceptible.

f) Type x A mixed reaction uredosori of variable size, mostly those of types 1 and 4.

FORMATION OF NEW RACES

All cereal rusts are heteroecious. An aecidial stage is never synthesized on a cereal host. As long as the fungus reproduces itself by means of uredospores, it breeds true except for the occasional appearance of a mutation. On the alternate hosts the sexual stage of the fungus is developed and this gives an opportunity for hybridization. This could result in the formation of heterozygous characteristics with differences in physiological specialization. When such an heterozygous race forms the basidiospores it segregates again and different recombinations may result again.

In this way the appearances of new races are recorded more in those regions where *Berberis* spp. plays an active role, as in the USA and Europe. On the other hand, in countries where *Berberis* spp. plays no part or little part in the life cycle, the number of races is much less, as in Australia, India and Kenya. In *P. gram in is triti* alone there are about 225 races in the USA, 9 in Australia, 11 (and 5 biotypes) in India and 5 'in Kenya. In other species of *Puccinia,* the number of physiological races is lesser: In *Puccinia recondita* there are 90 in the USA and 12 in India. In *Puccinia striiformis* there are 30 in the USA and 10 in India.

Biotypes

(a) A biotype is a population of individuals that are genetically seen. The descendants of a single non-sexually produced stem rust spore constitute a biotype. The biotype will remain totally pure unless a mutation occurs to cause genetic diversity. Most species of pathogenic fungi are now known to have many genotypes. The most closely related biotypes are grouped together to constitute a race. The closely related races' are similarly grouped into varieties and finally the varieties into species. (b) The larger groups (varieties) are recognized first and then smaller groups (races and biotypes) are easily distinguished by more refined methods. Biotypes show enough similarities to be placed within one race, but when different biotypes are studied very closely, they show minor but consistent differences in their effects on some differentials, proving that they are genetically pure, though different identities.

(c) Another instance may be given. Race 15 of wheat stem rust was discovered in 1918. It was a fairly homogenous rust for a long time and was identified as such. Later, some dissimilarities in isolates from different regions were noted. These differences were consistent enough and 15A and 15B were recognized as biotypes. Race 158 which was known to occur occasionally near barberry bushes became widespread in 1950 and attacked the well-known hitherto resistant varieties. .Besides rusts, specialization of parasitism is found in other fungi . (d) In smuts, as in rusts, races and biotypes are known to occur in different species. As a matter of fact, smuts offer a better material for the study of genetics. In smuts *(Ustilago zeae)* the chlamydospores undergo reduction division and the sporidia produced are haploid structures which can be grown into haploid mycelium in culture media and matings done as per plan. The colonies from haploid" sporidia are genetically pure and differ from one another in characters such as size, colour and surface patterns. It has now been known that in *Ustilago zeae* there are thousands of haploid biotypes within the species. When we talk of races in rusts, we consider dikaryotes, but in smuts, we consider haploid biotypes. Biotypes may differ from one another in one or more of all known characters, including pathogenicity. Similar though less elaborate work has been done on other smuts, for example, *Ustilago tritici, Ustilago nuda, Sphacelotheca sorghi, Ustilago avenae* and *Ustilago hordei.*

(e) Investigations in rusts and smuts are mutually complementary. In rusts only the binucleate uredial stage can be propagated clonally, while in smuts it is only the haplophase which can be propagated in this way. In both smuts and rusts three distinct phases correspond with the nuclear condition, the haplophase, the dicaryophase and the diplophase. In rusts, both dicaryophase and haplophase are parasitic in life, but in smuts, the haplophase is generally saprophytic. The dicaryophase, both in smuts as well as rusts, can be hybrid in character. This character is multiplied (in an equational manner) vegetatively through the uredospore stage. In many cases, rusts can go on repeating indefinitely from year to year. In smuts, such multiplication does not occur and sexuality is an indispensable 'process. Variability in smuts is, therefore, more common, and hence there is a very large number of races, but these races have to be expressed in terms of haploid strains and not in the dikaryotic phase, as in rusts. No sooner is the chlamydospore formed in smuts than the dicaryotic phase comes to an end and is not propagated clonally (by uredospore, like conidia) as in rusts.

Erysiphe graminis is also strongly specialized into physiological races.

ADAPTATION OF FUNGI TO VARIABLE HOSTS

The Gene-for-Gene Theory of Host Parasite Relations

Flor, in his studies on rust disease of flax *(Melampsora lini)*, came to the conclusion that for each gene that controls resistance in the host there is a specific and related gene in the pathogen that controls its virulence (pathogenicity). In other words, there is a genetic inter-relationship between the host and pathogen. This is the complementary gene-for-gene theory. The mechanism for the evolution of this relationship in a host parasite system is an automatic consequence of the distribution of new resistance genes by mutation and selection in the host, and virulence genes in the parasite population. The genetic definition of specificity embodies the so-called 'quadratic model' in which genes in the host are clearly defined in terms of resistance or susceptibility and the complementary genes in the pathogen in terms of virulence or avirulence.

The relationship between the host and parasite induces a

genetic polymorphism in both partners. This genetic variability, exhibiting a self regulating automatic mechanism, is the reason for the vicious circle that consists of the alternative prevalence of the resistant host and the virulent pathogen.

The gene-for-gene hypothesis was proposed by Flor as the simplest explanation of the results of studies on the inheritance of pathogenicity in the flax rust fungus *(Melampsora lini)*. In the varieties of flax *(Linum usitatissimum)* that have a gene to resist the avirulent parent race. F_2 cultures of the fungus segregate into monofactorial ratios. In the varieties that possess 2, 3 or 4 genes for resistance to the avirulent parent race, the segregation in the F_2 cultures is in bi, tri or tetrafactorial ratios.

Flor summarized the evidence for flax rust in the gene-for-gene hypothesis. This states that during their evolution, the host and parasite develop complementary gene systems so that "for each gene conditioning rust reaction in the host there is a specific gene conditioning pathogenicity in the parasite". The current status of the gene for gene concept has been recently reviewed by Flor. This hypothesis is applicable to most host-parasite systems in which resistance is conditioned by major (vertical resistance) genes and virulence increases in a step-wise manner. Hatchett and Gallun showed that the ability of specific races of Hessian fly to survive on certain varieties of wheat is controlled by genetic systems in the host and insect that are complementary.

The reactions of host plants in the presence of different pathogenic races are controlled by many genes, and within them, an even larger number of alleles exist to enable resistance.

Resistance in plants is often dominant over susceptibility to disease. It is generally found that resistance of plant cultivars to diseases is not a feature character. The reason for this phenomenon is to be found in the self regulating nature of genetic variability in host parasite relations. The evolution of new genes for virulence in the pathogens is the result of the selective process of resistant cultivars on populations of the pathogen which are genetically plastic.

Similarly, the genes for virulence in parasites promote the selection of genes for resistance in the host population. This automatic regulation of populations of the host and the parasite explains the fact that in many cases the utilization of resistant cultivars is very limited. For example, as a rule, rust resistance is lost in three to seven years. This is valid, first of all, for pure line cultivars of self-pollinated plant species that is, for wheat which suffers from attack by a new rust race much more heavily than the crosspollinated plants (*e.g.* cultivars of corn). Cultivars of cross-pollinated plants are composed of a mixture of different lines (biotypes). This means that some biotypes will be resistant in the host-population to the selection effect (attack) of the virulent gene of the parasite.

However, there are very few, if any, resistant biotypes in the susceptible population of pure line cultivars, since they are largely eliminated during the breeding process which produces a high degree of homozygosity. Thus, pure lines will be highly vulnerable to attack by the parasite containing genes for virulence. On the other hand, in cross pollinated plants, the buffering capacity of the genetic pool is relatively high as compared to the self-pollinated ones. In recent years plant breeders have tried to incorporate greater variability into self-pollinated plants such as wheat by developing multiline cultivars. The various lines in the multiline cultivars have different gene-for-gene systems against the virulent genes of rust fungi. By continually manipulating intra-varietal heterogenity within available resistance genes, it should be possible to provide a stable control of widespread pathogen inoculum potential.

The advantage of using multiline cultivars, according to Sadasivan, is that they enable quick synthesizing of well buffered 'horizontal resistance' cultivars (HR). They are capable of utilizing, unlike pureline cultivars, several resistance genes at the same locus. The useful life of a resistant gene can be extended indefinitely and the breeding programme can be cut to size by this method.

There is an even chance for the cultivars to stabilize themselves in a given environment. This process helps to distribute developed varieties on a wide geographical scale without any risk of homogenizing the pathogen population. The multiline approach is supposed to be expensive, agronomically conservative and a breeding ground for new races, possibly even a super race. Another drawback is that because their variability is synthetic it is polarized.

The resistant cultivars of cross pollinated plants remain resistant for a considerable time. One such example, according to Borlaug (1965), is the relative stability of resistance in corn cultivars to the rusts, *Puccinia sorghi'* and *Puccinia polyspora.* These two rusts of corn are much less important in limiting corn production than the rusts of wheat. The crosspollinated corn usually possesses sufficient variability, that is, sufficient resistant lines are present in the population of the cultivar to resist the selective action of virulent genes of new rust races. This means that the development of rust epidemic, in cross-pollinated corn is almost impossible.

RESISTANCE AND SUSCEPTIBILITY

A plant is either immune to a pathogen, or it may show differ degrees of resistance ranging from immunity to complete susceptibility. Resistance may be conditioned by a number of internal and external factors which operate to reduce the chance and degree of infection. There are several varieties of resistance, they may be physical or chemical, it may depend upon the lack of a nutrient or a substance that the parasite needs for development, or it may be due to a toxic or repellent substance that is either preformed or formed only in response to infection. Irrespective of whether the morphological defense mechanism or biochemical defense mechanism is involved in the resistance of a particular host-parasite combination, it is ultimately the genetic make-up of the 'two that determines whether a particular host will be susceptible or resistant under a given environmental condition. The mechanism by which genes control the physiological processes that Lead to disease resistance or susceptibility are not yet clear, but they presumably are no different from the mechanisms controlling any other physiological process in a living organism. Under the influence of the genetic material (DNA), a specific protein, which may be a structural protein or an enzyme, may be produced. The produced enzyme, of course under certain conditions, will participate or initiate biochemical reactions related to one or the other of the cellular processes, and may result in the production of a certain morphological characteristic or accumulation of certain chemical substance.

Unlike higher animals, plants as far as we know, have not evolved a central coordinated mechanism to deal with an invading micro-organism. There is no substantial evidence of a highly specialized system that routinely recognizes foreign entities and responds to them in a uniform and predictable way by antibody formation. The response of the plant is highly individualized and appears to be localized sprimarily in the cells directly under invasion. To quote Daly, "the confrontation is 'eyeball to eyeball' between the host and the parasite and the outcome is decided at the site of invasion".

POSSIBILITES FOR RESISTANCE

I. PREFORMED (PASSIVE BARRIERS)

A. MORPHOLOGICAL

B. CHEMICAL

II. INDUCED (ACTIVE) BARRIERS

PATHOGEN

↓

INDUCER

↓

BIOCHEMICAL

NEW MORPHOLOGICAL BARRIERS ← ACTIVATION → INSENSTIVITY TO INDUCER

PATHOGEN INHIBITORS

INDUCER DESTRUCTION

Fig. Passive (pre-formed) and active (nduced) barriers to infection by plant pathogen.

Historically, plant pathologists have been concerned with the nature of disease resistance because of its obvious economic importance. In the early part of this century, considerable effort was spend in looking for morphological or chemical characteristics of uninfected plants which explain their resistance to certain diseases. Figure 7.a indicates the possibilities for resistance.

Although pre-formed or passive barriers to infection undoubtedly exist, two other scientific developments gradually caused a shift to other lines of thinking. First, there was the recognition of the existence of pathogenic races within a fungal or bacterial species, races which differed only in their ability to infect certain varieties of a host. The varieties show no obvious morphological or biochemical differences. In the case of wheat rust, there are specific dominant genes for host resistance and corresponding recessive alleles for virulence. If all the genes for resistance are absent from the host, the host is completely susceptible to all pathogenic strains. A specific gene for resistance results in incompatibility to all the races except those which have the recessive allele for virulence against the host gene.

Figure shows the expected results for a host-parasite system with three gene pairs. With 3 loci, there are 2^n phenotypes, and thus, pathogenic races are possible; with 8 loci, there could be 250 races. In the case of wheat stem rust, some 17 loci are known.

LOCI AND RACES (after Person)

PATHOGEN				H_1	H_2	H_3	H_4	H_5	H_6	H_7	H_8
GENES FOR					A			A	A		A
VIRULENCE						B		B		B	B
(n=3)							C		C	C	C
P_1				S							
P_2	a			S	S						
P_3		b		S		S					
P_4			c	S			S				
P_5	a	b		S				S			
P_6	a		c	S					S		
P_7		b	c	S						S	
P_8	a	b	c	S	S	S	S	S	S	S	S

LOCI=n PHENOTYPES=2^n

8 LOCI= 250 RACES

9 LOCI= 500 RACES

Fig. The genotypes and phenotypes in a host parasite

The concept of resistance mechanisms, either passive or induced, has been a main theme in all our thinking. The genetic models discussed are based on the assumption that the genes in the plant or pathogen, or both, control the production of substances which determine the nature of disease reactions. Flors-gene-for gene concept implies that resistance is an induced responsed that requires the production of an inducing substance by an active gene for avirulence in the pathogen. If the gene for avirulence is inactive or absent, passive susceptibility will be the result.

Selective pathotoxins, cell wall degrading enzymes, specific inducers of resistance, presence or absence of common antigens and phytoalexins are involved in the specificity of plant pathogens to their hosts.

4

Study of Pathogenesis

CONTACT OF PLANT PATHOGENS WITH HOST PLANTS

Contact

Pathogenesis caused by an infectious agent starts as the pathogen comes into contact with any portion of a plant. Most of the bacterial and fungal pathogens come into contact with their hosts accidently. Many pathogens are carried by insects, nematodes, and other vectors. Motile propagules of fungal pathogens are attracted to root exudates such as the zoospores of many plant pathogens. Attraction and accumulation of zoospores to roots in the soil just behind the root tip has been demonstrated.

Accumulation of zoospores is the result of a sequence of responses. Till now the precist nature of the start is not entirely clear. It is possible that several responses and not just one are involved in accumulation of 300 spares.

Interesting research by Kerr and Flentje showed that the hyphae of the fungus *Rhizoctonia solani* accumulate on the outside of cellophane bags in contact with the roots, indicating that root exudates induce the formation of infection cushion by the fungus.

PENETRATION AND INFECTION BY PATHOGENS

When a pathogen comes in contact with a host plant, many important factors play part in deciding weather infection will occur or not. For example temperature, moisture, susceptibility of plant tissues, effects of other micro organisms present the Pathogenesis is a chain of events by which disease development takes place. These chain of events can be lighed as:

A. Prepenetration.

B. Penetration.

C. Postpenetration or further colonization of infected plant by the pathogen.

PREPENETRATION OF PATHOGEN

In fungal pathogens, prepenetration includes germination and growth of the germ tube on the surface of the host plant. Germination is of spore the change from a low metabolic rate to a high metabolic rate and involves a change from near-dormancy to high activity; for this an energy source is needed such as a carbohydrate or fat reserve in the propagule. Fungal invasion is chiefly by germ tubes or structures derived from them. By two ways Hyphae of *Rhizoctonia solani* often aggregate to form an infection cushion from which multiple penetrations occur by means of appressoria and penetration pegs. This is shown in figure. The other example of penetration is exemplified by the penetration of intact root periderm by the rhizomorphs of *Armillarly mellea,* which is shown in figure. The bacteria and viruses enter in various ways. According to the nature of the spore and the environmental conditions, germination of fungal spores occurs in various ways.

Sporangia and oospores of Phycomycetous fungi produce zoospores in wet conditions and germ tubes in dry conditions, Teliospores of Uredinales and Ustilaginales normally germinate by producing a promycelium on which basidiospores (sporidia) are borne. Budding and fission are characteristics of some yeasts and bacteria.

Some spores germinate easily on release and sometime even before release, while resting spores have a dormancy period during which they do not normally germinate. These can be due to chemical and biological factors. The behaviour of germ tubes growing on a plant surface depends on many factors. In order to get food, plant pathogens have to make a close connection with their hosts. The pathogen invades plants through natural openings (stomata, lenticels, nectarthodes, hydathodes), through artificial openings (natural or due to insects or nematodes), or by penetration. While some enter in one way only, others enter in more than one way.

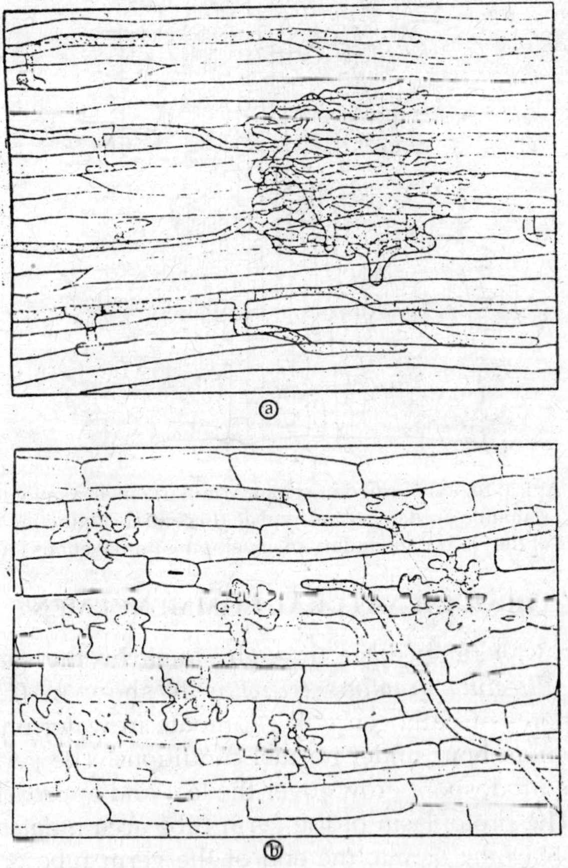

fig. Formation of infection cushion by crucifer isolate of R. Solani on radish hypocotyl (b) production of lobate appressoria by solanaceous isolate of R. solani on tomato hypocotyl.

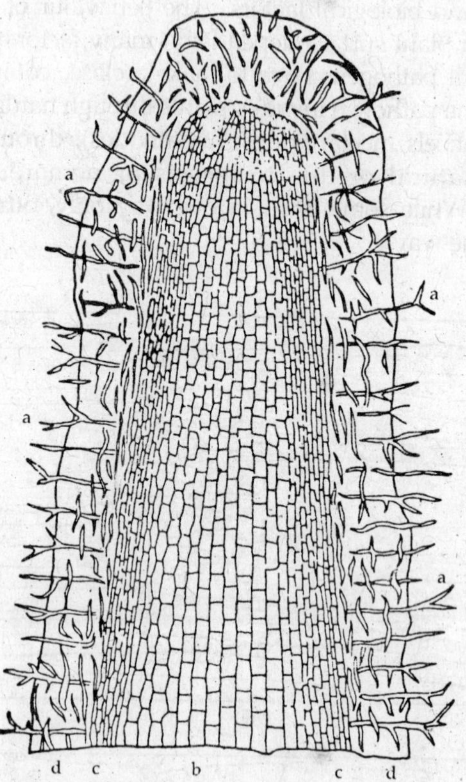

Figure. Apical region of rhizomorph of Armillaria mellea in logitudinal section: (a) mantle of filamentous hyphae (b) central conducting cells (c) cells of the rind (d) boundary of enveloping muilagenous layer.

ENTRY THROUGH NATURAL COMMENCEMENT

The way fungi enter through stoma by the stem rust fungus *Pllccinia graminis tritici* is is show in (Figure). Uredospores of pathogen will germinate in water on a leaf surface in an hour under normal conditions. The germ tube from the uredospore grows over the leaf and approaches the stoma. The protoplasm of the germ tube accumulates at the tip. On reaching stoma, the end of the germ tube grows to form an appressorium in the stomatal aperture. The empty germ tube is cut off by the formation of a cross wall. From the appressorium a blade-like wedge grows through the stomatal slits and swells inside to form a substomatal vesicle into which the contents of the appressorium pass. One or more hyphae grow from the vesicle towards the cell lining the substomatal cavity and after making contact proceed to penetrate the cell wall. After penetration, a haustorium is produced in the cell. In diseases caused by obligate parasites and facultative saprophytes and in a few caused by facultative parasites, there is no massive penetration of the cell cavities of the host by ordinary hyphae of the parasite.

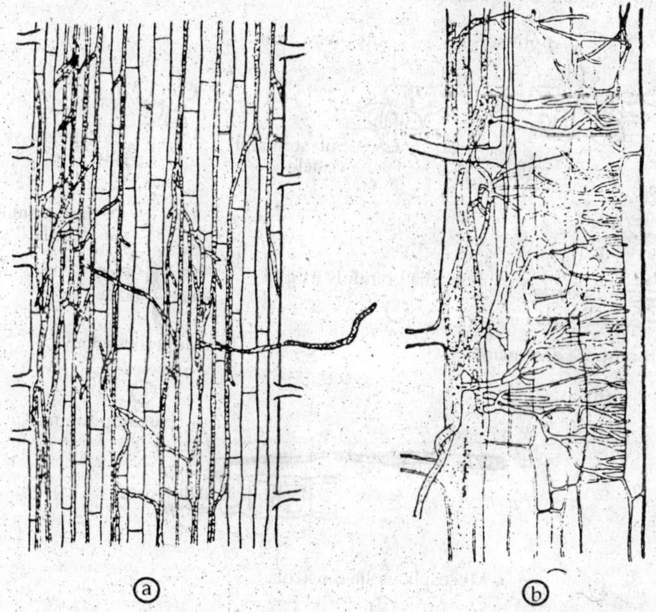

Figure. Ectotrophic infection of *Ophiobolus graminis:* (a) in surface view (b) in longitudinal section

Instead, the cell wall is penetrated by short branches which usually grow to a limited extent in the lumen of the cell before their ends are transformed into structures called haustoria, the forms of which may be simple or relatively complex depending on different parasites. Haustoria are the main source of metabolic exchange between the host and parasite.

Where as zoospores of a species of *Phytophthora* infecting fennel *(Foeniculum vulgare),* always enter by stomata while

Phytophthora infestans always enter plants directly hut at times through stomata after the formation of an appressorium. In *Pseudoperonospora cubensis* (Figure) the zoospores swim along the boundary of the epidermal cells towards the stomata and encyst above the line during the guard and epidermal cells. Hyphae prierce the stomatal opening and expand to form a substomatal sac from which in turn other hyphae may grow to form haustoria in the next cells of the leaves. *Mycosphaerella musicola* (Figure) responsible for Sighatoka disease of bananas, enters leaves in a similar manner as do the rusts.

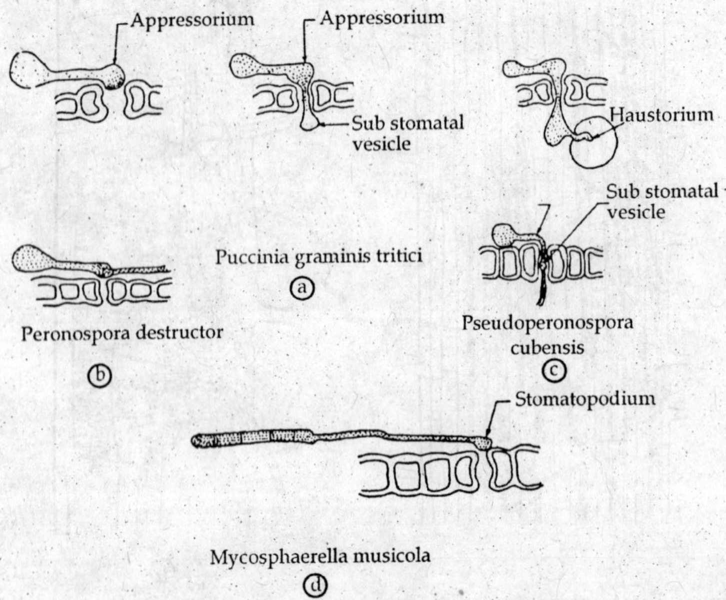

figure. Penetration through natural openings.

The germ tube from a conidium grows slowly over the leaves for a few days before its entrance. This shows that it can withstand periods of drought/dryness. A small structure called a stomatopodium is formed over the apestore of a stoma. A hypha then grows from it which emerges into the substomatal chamber and expands to form a sac, which in turn gives rise to hyphae which encroach palisade tissues sac. Leaf infecting fungi-imperfecti generally invade through stomata. Some of the examples are *Cladosporium fulvum* infecting tomato; *Cladosporium cucumerinum* infecting cucumber and *Stemphylium solani* infecting tomato.

The method of invasion into plants by *Sclerotinia cinerea* which causes stone fruits to decay depends on the host species infected. It invades plums through the stomata, apricots through the stomata or cuticle, peaches through hair sockets and cherries and nectarines through the cuticle.

Stomata are the most important or the only route of entry for many parasitic bacteria such as *Pseudomonas tabaci* on tobacco, *Erwinia amylovora* infecting apples and pears, *Xanthomonas phaseoli* infecting *Phaseolus* spp., *Xanthomonas malvacearum* infecting cotton. Bacteria enter stomata only by films water that stretch out from outside through the stomatal aperture into the inside of the shoots. Due to this water content achieve importance to check if they will get infected. In case of citrus fruits, the conformation of the cuticle around the stoma either prevents or allows the passage of water droplet containing the bacterial pathogen *Pseudomonas citri*.

Lenticels

In many dicotyledonous plants, a defensive layer grows as the epidermis becomes redundant which is called Periderm periderm also grows some common methods of entrance are on surfaces exposed by abscission of plant parts or on surfaces produced by the wounding of tissues: It also develops on the surface of storage tissues, such as potato tubers.

a) Streptomyces scabics, the cause of the common scab of potato, invade young developing tubers through stomata or lenticels.

b) *Oospora pustulans* which causes the skin scab of potato tubers also invades through lenticels.

c) *Penicillium expansum* which is cause for storage rot of apples and *Gloeosporium perennans*, the cause of another storage rot, also invade through stomata or lenticels:

Nectria galligena, the cause of apple canker, can invade twigs through lenticels.

Nectarthodes, Hydathodes and Trichomes

Erwinia amylovora invades by the way of nectaries of apple and pear flowers. In pears, nectaries occur in saucer-shaped tissues between the point of growth of the style and stamens.

Small drops of nectar are discharged through stoma-like nectarthodes at the base of depressions on the surface of the nectaries. Within 48 hours of adding suspensions to open pear blossoms, the bacteria are found locally in great numbers beneath the nectarthodes. Hydathodes are structures through which water is exuded from the interior of leaves, usually more or less directly from bundles of terminal tracheids which end in thin walled parenchyma known as epithem. In high humidity bacteria amass in water from hydathodes and are dragged into epithem and tracheids, as the water draw back when conditions reduce water congestion.

DIRECT PENETRATION

Direct encroachment by pathogens of plants has been known for many years. De Bary was the first to study critically the nature of direct penetration by *Sclerotinia sclerotiorum.* He emphasis on the relationship of nutrients in the infection drop. The physicochemical characteristics of cuticles and cell walls are important factors in direct penetration.

Structure of Cell Walls and Cuticles

The external layers binding the intercellular spaces and in between the adjacent cells is made up of largely of pectic substances, particularly calcium and magnesium pectates and can be broken down by the fungal and bacterial pathogens which produce pectolytic enzymes. Proteins also occur in the middle lamella. Within this is a primary cell wall formed mainly of cellulose embedded in a matrix of pectic substance and hemicelluloses. The cellulose may be present as micellae. Other substances, such as proteins, lignin, suberin and cutin are also prevent the primary cell wall. In some tissues, a secondary cell wall consisting chiefly of cellulose, is present within the primary wall. The cuticle is a non-cellular membrane which covers most of the aerial parts of the higher plants. The outer waxy layer is hydrophobic and largely impermeable to water but the inner layer can absorb moisture, and swell and for the wax aggregates apart. The innermost layer of the cuticle passes gradually into a predominantly pectic layer which in turn merges into the largely celluose outer walls of the epidermal cells.

The uninjured epidermis which is covered by a well defined layer of cuticle is invaded directly by a mechanical process. In most cases, penetration by the formation of a special structure, namely, penetration hypha or appressorium which puts a mechanical pressure on the intact wall of the host. Infection peg, arising from the appressorium, pierces the host wall to cause infection. The invasion of the cuticle is regarded by many as purely mechanical, although there is evidence of the chemical break down *of* the cuticle. Many *root* pathogens amass on the surface of roots and hypocoty and other subaerial parts and penetrate directly. The penetration mayor may not be preceded by the growth *of* an appressorium.

The primary infection of wheat and barley by *P. graminis* consist is composed of a number of morphologically identifiable stages of development, that is, spore germination formation of appressoria, penetration into host cells, formation of haustoria in host cells and formation of elongating secondary and tertiary infections. The wax layer of the cuticle surface is a major factor for the formation of appressoria.

In direct invasion of a leaf, the invading hypha which arises from the underside of an appressorium has to pass through the cuticle and the underlined tissue arid sometimes through the thickened outer wall of the epidermal cell, although some time passes between the epidermal cells. In many fungi this process has been studied, including *Cladosporium carophilum* [figure]. The smut fungi usually invades directly into young ovaries, shoot buds, young seedlings or young tissues.

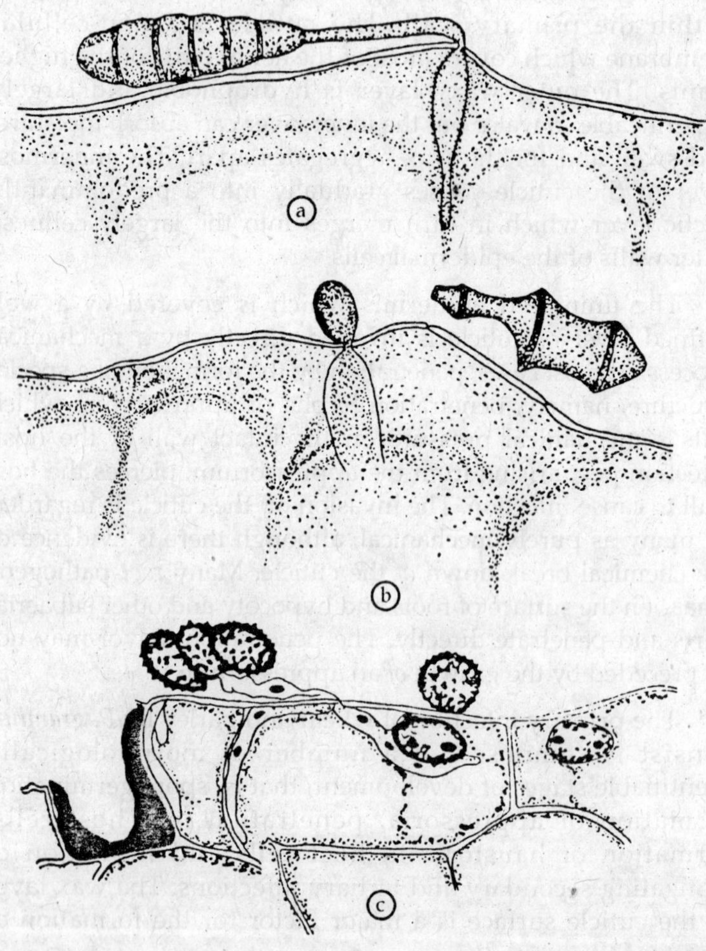

Figure. Direct penetration of the cuticle. Penetration of almond leaves by germ tubes of *Cladosporium carpophilum,* the penetration hypha passing (a) between epidermal cells (b) passing into the epidermal cell (c) direct penetration of corn cells by the smut fungus, *Ustilago maydis,* the germ tube penetrating directly through an epidermal cell and making contact with host nucleus.

In the case of infection due to *Erysiphe graminis* on wheat leaves, a fine stylar process from the germ tube penetrates the cuticle, and in its passage through the cellulose layer it is preceded by a local thickening of the layer into a papilla which it eventually pierces at the apex. Inside the host cell it becomes dilated into a branched haustorium (Figure).

The starting formation and action of appressoria are intriusic parts of the infection process of many parasitic fungi. The formation of appressoria may be essential for infection, while sometimes it may be unnecessary. The term appressorium was introduced by Frank for "spore-like organisms" formed on the germ tubes of *Colletotrichum lindemuthianum, Polystigma rubrum* and *Fusicladium tremulae.*

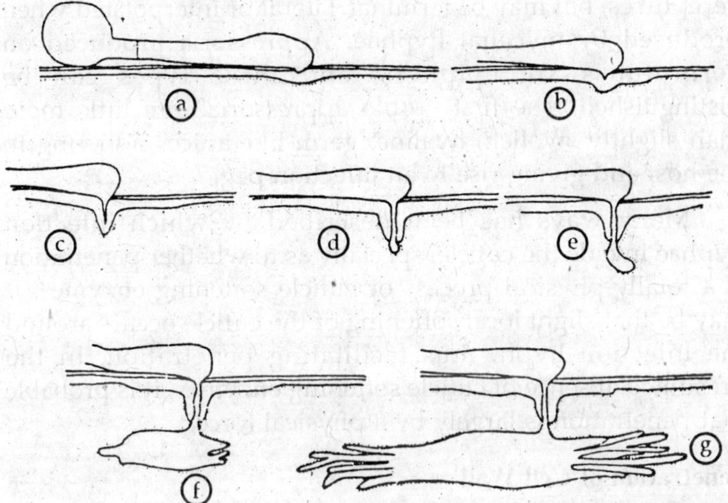

Figure. Germination of conidia of *Erysiphe graminis* on the epidermis of wheat: (a) germination of conidium (b) a fine stylar process from the germ tube (or from a small subapical appressorium) pierces the cuticle (c) local thickening of the layer to form a papilla (d) papilla eventually pierces at the apex (e) dilation inside the host cell (f) development of a lobed haustorium (g) branched haustorium and invagination of the lining layer of cytoplasm (x2000)

He believed that appressoria are adhesive discs attaching a parasitic fungus to its host during the early stages of infection.

Appressoria may be simple or complex/compound. Some fungi produce appressoria at different stages of growth; some produce more than one kind. The dematiaceous epiphytic parasites mostly produce appressoria on germ tubes identical to those produced latter on hyphae. In the members belonging to Erysiphaceae, most species produce appressoria which are different in form on hyphae than those on germ tubes, each type being morphologically constant for a species.

Appressoria can be divided into two groups a) simple appressoria made up of single modified cell and b) compound appressoria made up of many cells.

The main aim of appressoria is to penetrate host and produce infection peg. Appressoria produced by germinating uredospores of rusts and *Mycosphaerella citri* penetrate through stomata. Simple appressoria occur terminally on germ tubes, but may be terminal, lateral or interpolated when produced by mycelial hyphae. Appressoria produced on germ tubes are epiphytic and three types can be distinguished. The first "proto appressoria" are little more than slightly swollen, hyaline, germ-like apices adhering to the host and giving rise to an infection peg.

Many ways has been described by which infection hyphae invade the cuticle, specially as to whether penetration is a totally physical process or cuticle softening enzymes. It may be that slight local softening of the cuticle occurs around the infection hyph, thus facilitating penetration. In the presumed absence of cuticle softening enzymes, it is probable that penetration is largely by a physical force.

Penetration of Cell Wall

After passing the cuticle, the infection hypha has to grow through the outer wall of the epidermal cells which is made up of a matrix of cellulose microfibrils in a matrix principally composed of pectic materials. The infection hyphae of some fungi pass between the radial walls of the epidermal cells without entering them. The penetration of the walls is largely due to action of enzymes. The infection hyphae grow quickly through the softened materials. Penetration of the cell walls may occur within a minute and is probably a combination of physical pressure and enzymic softening. Swelling of the invaded cell wall and of walls some distance away may occur due to the diffusion of enzymes. Swelling may be of many type like, localize obliterating etc.

Infection can be as easily caused by mycelial hyphae of

many fungi as on germ tubes from spores.

The germ tubes of aeciospores of *Puccinia menthae* (mint rust) produce infection structures resembling appressoria when expose to thymol at its own saturated vapour pressure on glass but not on glass in the absence of thymol. Thymol also reduces spore germination and elongation of germ tubes. It is thought that it acts as a morphogenetic switch in bringing about the formation of appressoria.

In the case of infection by *Rhizoctonia solani,* infection pads are formed on the roots or hypocotyls of the infected plants. Multicellular infection cushions are produced from a branch hypha in which normal elongation has ceased but prolific side branching now takes place. Strains of *Rhizoctcnia solani,* non-pathogenic to seedlings of a particular species failed to become attached to the hypocoty and organize the infection cushions.

ENTRY THROUGH WOUNDS

Many low grade pathogens are not able to pierce the host surface in the absence of a wound. However, once in the plant, they often to cause serious disease symptoms. Wounds may be natural or they may be caused by man or other factors, such as insect bites. Many horticultural and agricultural practices involve accidental or even deliberate wounding. Other natural wounds are those that occur due to the growth of laterals from main roots. This is a means of invasion for the tobacco root-rot fungus *Thielaviopsis basicola* and for *Fusarium solani* f. *phaseoli* into bean roots. Many important forest pathogens also enter through wounds. Man's activities are responsible for many of the wounds through which fungal penetration occurs. Many fruit and vegetable products are quite ripe when they are ready for marketing. In this condition they are most susceptible to wounds by rough and careless handling.

An unusual type of entry of a variety of micro-organisms occurs through wounds in flax and maize seeds. The damage is mechanical during or after harvesting. Somewhat similar is the infection of wheat seedlings by *Cephalosporium gramineum* which enters through wounds caused by the breaking of the pericarp and coleorhiza as the root emerges on germination. Wounds caused by insects are important in many diseases In some diseases, the relationship between the insect and parasite is non-specific while in others the relationship may be more specific, the distribution of the disease coinciding with the appearance of a particular insect species.

Among the diseases caused by bacteria, vascular wilt of cucurbits is one where the relationship is highly specific. This wilt is caused by *Erwinia tracheiphila* which overwinters in hibernating adults of the cucumber beetles *Diabrotica vittata* and *D. duodecimpunctata.* When adults feed on plants, they introduce the bacterium into the vascular system, where it spreads to cause a typical vascular disease.

There is a similar relation in bacterial wilt and Stewart's leaf blight of corn caused by *Xanthomonas stewartii.* The bacteria overwinter in the intestinal tracts of insects, in this case the flea beetles and plants become infected when beetles begin to feed in spring.

In another disease the relationship is more casual, for example, in fire blight of apple and pear leaves caused by *Irwinia amylovora* where the bacteria overwinter in cankers. With the return of growing conditions in spring, bacteria multiply in the cankers and are carried to the blossoms by insects.

A large number of fungal diseases are distributed by insects and are instrumental in their entry into the host. In the ergot disease of rye caused by *Claviceps purpurea* conidia are disseminated by species of *Diptera* and *Coleoptera.* In oak wilt caused by *Ceratocystis fagacearum* the relationship between the fungus and insects is also relatively non-specific.

A variety of insects also cause feeding wounds, through which species of *Sclerotinia* invade stone and pome fruits.

WOUNDS CAUSED BY OTHER FUNGI

Resistance to infection can be used by other fungi. Tubers which are highly resistant to *Phytophthora infestans* become easy pray when the parasite enters through wounds made by *Spongospora subterranea.* The apple canker fungus *Nectria galligena* may enter through wounds on a stem caused by *Venturia inaequalis.*

WOUNDS CAUSED BY NEMATODES

It is well established that infection of plants by some bacteria and fungi is dependent increased by damage caused by nematodes. Wilt of solanaceous plants caused by *Pseudomonas solanacearum* is done by *Meloidogyne incognita* var. *acrita.* In the case of yellow ear rot of wheat the cause of which is the bacterium *Corynebacterium tritici,* another nematode *tritici* plays an important part in the infection process. Several vascul wilt pathogens enter into the vascular plants through the injuries caused by the attack nematodes.

ENTRY BY ROOT HAIRS

Root hairs are particularly susciptible to invasion by pathogens. Among the pathogen which can enter through root hairs, are *Plasmodiophora brassicae* (finger and toe disease crucifers), F. oxysprum f. sp. *conglutinans* (cabbage yellow). The method of entry by *Plasmodiophora brassicae* appears to be quite distinguished. This process involves several unusual structural features, Zoospores of the pathogen makecyse on the wall of a root hair. The contents of the zoospore are injected into the host cell by a fine puncture or hole made in the cell wall of the host, The puncture is the result of the production of a bullet-shaped structure which suddenly Pierces its way from within the cyst through the wall into the interior of the root hair cell. This is an example of mechanical penetration and the actual infection process takes only a few seconds . Another interesting example of entry through root haris is by the root nodule bacterium *(Rhizobium)* or leguminosae which enters through root hairs in the form of an infection thread.

ENTRY THROUGH BUDS

Some rust fungi, including *Uromyces pisi* enter by the way unfolding buds. *Taphrina cerasi* attacks cherry buds and causes them to develop into typical witches brooms. The peach leaf curl fungus *T. deformans* enters its host though the unfolding leaf-buds.

An unusual mode of entry is that of the pathogen *F oxysporum* f. sp. *pisi* which enters through cotyledonary bundles which are exposed when cotyledons are rotted by soil sapro- phytes (Hepple, 1963).

PROCESS TAKING PLACE IN HOST TISSUE

After penetration there are very wide variety in the subsequent method of development of the host plant by pathogens of host tissues. In many pathogens, such as powdery mildews, the growth of the pathogen is very superficial parasitic fungi infect and accumulate in the interior of the plant and also produce an external aerial mycelium which grows superficially and enters the host repeatedly, for example, *Corticium solani* responsible for stem canker of potato and *Ophiobolus graminis* responsible for take all diseases of wheat. Many pathogcns colonize the parenchyma of mesophyll and cortex; their further spread into the vascular bundles is prevented by the endodermis, although there may be an entry into the pith. In *Taphrina* the hyphae are entirely intercellular, obtaining nutrients through the memb- fanes of the adjacent cells and this in direct method perhaps results in some degree of systemic development. The hyphae of less specialized pathogens are often intracellular, progressing with branches from cell to cell until the tissue is more or less completely colonized. Toxins and enzymes may be secreted and bring about rapid softening and necrosis of tissues ahead of the pathogen. Some parasites enter living cells, which however, may die soon after penetration. In diseases caused by obligate parasites and facultative saprophytes and in few caused by facultative parasites, the cell wall is penetrated by short branches which usually grow to a limited extent in the lumen of the cell before their ends are transformed into structures which exit in simple to complex form in parasite. These structures are haustoria, whose main function is absorptive rare the main chance is of metabolic interchange between host & parasite Haustoria almost always occur in parenchyma but in most powdery mildews they are confined to epidermal cells.

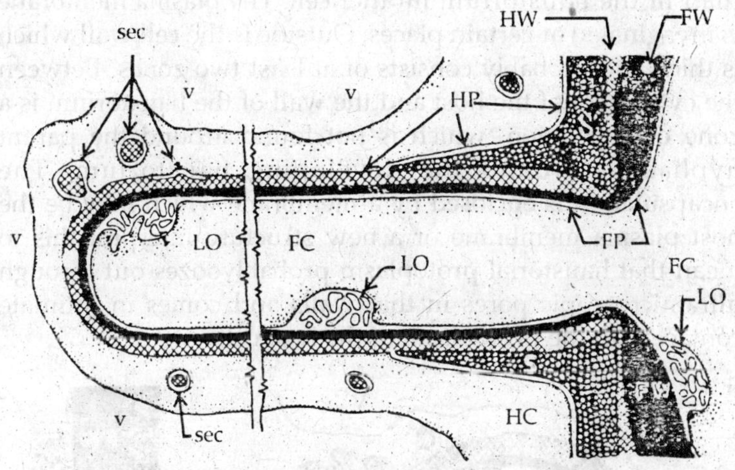

Figure Diagram of the structure of *peronospora manshurica infecting Glycine max* as revealed by electron microscopy. The cytoplasm (FC) of the fungus is bounded by plasma-membrane (FP), lomasomes (LO) and cell wall (FW). The host cell shows vauole (V), cytoplasm (HC), plasma membrane (HP), and cell wall (HW), ending in sheath (S). A zone of apposition (Z) separates haustorium from host plasma memorane. Invaginations (sec) of host plasma membrane and vesicular host cytoplasm suggest secretion by host cell. (Adapted from G.A. peyton and C.C. Bowen, *Am.J.Bot., 50: 787-797, 1963.)*

The haustorium originates as a slight swelling on the inside wall of the haustorium mother cell and it becomes closely attached to the host cell wall at the point it is to pierce. Here is formed a conical gap wide at the base and tapering at the apex, through which the haustorium entry. As it emerges, it is bounded only by a membrane continuous with that of the haustorium mother cell. As the haustorium is not sucrounded by a cell wall at this time, suggests that penetration is basically a chemical and not a mechanical process and that the haustorium mother cells produce substances that dissolve both their own walls and those of the host cells.

After penetration, the haustorium enlarge to form a neck around which a wall is formed. The apex of the neck expands to the characteristic final structure. There is some evidence of the invagination of host plasma membrane by the developing haustorium. The mature haustorium containing nuclei, mitochondria and endoplasmic reticulum is more complex than in the haustorium mother cell. The plasma membrane is invaginated at certain places. Outside is the cell wall which is thick and probably consists of at least two zones. Between the cytoplasm of the host and the wall of the haustorium is a zone of apposition, which is not found around the parent hyphae or around the developing haustorium. The encapsulation is enclosed by a membrane which may be the host plasma membrane or a new structure interpret this to mean that haustorial protoplasm probablyoozes out through ultramicroscopic pores in that walls and comes in intimate contact with the protoplasm of the host cell.

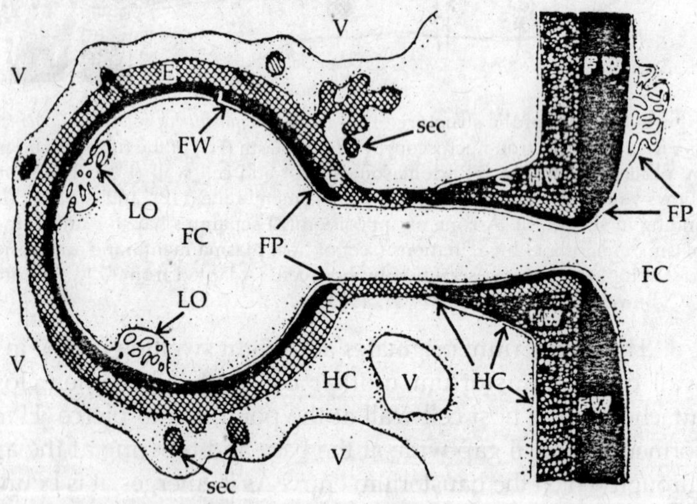

Figure. Diagram of structure of haustorium of *Albugo candida* infecting rophanus sativus as revealed by electron microscopy. Cytoplasm (FC) is bounded by plasma membrane (FP), lomasomes (LO) and cell wall (FW). The host cell shows vacuole (V), cytoplasm (HC), plasma membrane (HP), cell wall (HW) which ends in a sheath (S). An encapsulation (E) separates haustorium from host. Invaginations (sec) of host plasma-membrane my be secretory sites. (Redrawn after modification from J.D. Berlin and C.C. Bowen, *Am. J. Bot.*, 51: 445-452,1964.)

At some points, the surface of the encapsulation is broken down by small vesicles which invaginates into the host cytoplasm and which contain particulates similar to those of encapsulation. These vesicles or secretory bodies are formed in the host cytoplasm, which then move to the encapsulation and discharge their contents.

Encapsulation is also well developed in downy mildews. This is illustrated in Pic.c (viii) which shows diagrammatically the complex of host cell wall and haustoria for *Peronospora manshurica* on leaves of soybean and for *Albugo candida* on leaves of radish. [figure]

The basic structure is similar in both. The only main difference is the absence fungal cell wall around part of the neck below the head of the haustorium and the absence of nuclei *Albugo candida.*

5

Host Response to Infection

PERMEABILITY CHANGES

Before discussing permeability changes in diseased plants, It is necessary to first study the structure and function of membranes. The chemical structure of membranes shows that proteins and lipids are the major in gredients, making up 90-98% of the total dry mass. The rest are carbohydrates which are attached to proteins, lipids and inorganic ions. According to Robertson (1959), all biological membranes have a bimolecular lipid layer with the polar ends of the lipid molecules oriented outward and coated with monomolecular layers of protein in the extended configuration. Several models have been proposed after this but the most accepted view is the one by Hendle. According to this, the minimum structural requirement of the unit membrane is a lipid bilayer with polar regions both outside and inside. The lipid bilayer can be interrupted by protein pores or contain various granular or micellar inclusions and still provide a hydrophobic core. To remain alive in different environments, cells must regulate the kinds and quantities of substances they take in and release. Membrane permeability plays a vital role in this inward and outward movement of the substances.

Among the several illuers there that a sick plant shows some are primary while others are the secondary effects of the first. Regardless of the disease type or the nature of the pathogenic agent, changes in permeability shows in all diseased tissues which have been examined. Changes in permeability have been detected before any other symptom of the disease appeared or in advance of the pathogen This, plus the fact that changes in permeability can be caused by a variety of biological, chemical or physical agents, led to the suggestion that a change in cell permeability might be the initial event in pathogenesis.

Thatcher made the first advanced study of the extent of permeability changes in plant diseases, that is, rusts, powdery mildews, soft rots, dry rots and wilts and concluded that:

a. Cells of the fungus pathogens have a higher osmotic pressure than those of their hosts.

b. Altered permeability of the host cells is consistently associated with disease.

c. For rusts, which were most extensively studied an increase in host cell permeability characterizes susceptibility, while decreased permeability results in resistance.

Permeability changes also occur in diseases caused by bacteria and viruses. Membranes become leaky, due to which cells fail to gather salts and other materials. The milieu interior is disturbed and an ionic imbalance sets in. Changed permeability has become characteristic of diseased plant tissues. Cook and Stall observed a loss of electrolytes from the leaves of susceptible *Capsicum annuum* infected *with Xanthomonas vesicatoria.* and observed a greater and quicker loss of electrolytes from hypersensitive tissues. Ghabriel and Pirone demonstrated a causal relationship between permeability changes in root tissues add wilting in tobacco pepper plants *(Capsicum frutescens)* infected with tabacco etch virus.

A resembling ionic imbalance has beeu reported in *Verticillium* wilt of cotton and disease of rice by Dube. All these scicatists reported K + and Ca'' to be lost maximally which, when supplied externally, provides resistance. Calcium determine overall thickness of the unit membrane; and of all elements. only calcium deficiency results in membrane disruption. Bateman and Millar. Beckman and Kuc related calcium to resistance in plants infected by *Rhizoctonia solani.* Membrane damage has been observed in the electron-photomicrographs of infected tissues. Hall and Mount and Bateman have reported that endoPGTE, a pectic enzymes destroys plasma membranes and causes the leakage of ions. The breaking of plasma membrane occurs at plasmodesmata due to excessive in flow of water from hypotonic solution into the host cells.

Spalding and Abdul-Baki have shown that maceration and cell death were closely when potato discs were treated with a purified pectinmethyl trans-eliminase produced by *Penicillium expansum.* Tissues exposed to pectic enzymes undergo changes in permeability before cells are killed; the relationship among these occurancy is not clear at all. If tissues are plasmolyzed in hypertonic solutions, cell death permeability. changes can be greatly delayed during exposure to pectic enzymes, but such conditions have little effect on the maceration.

The substitution of pathogens by toxins has greatly helped in understanding the various steps in pathogenesis. Linskens treated tomato tissues with fusaric acid and lycomarasmin. Ions of Na+, K+, Ca++, as well as free amino acids were detected in leaf exudates after treating there with fusariç acid and lycomarasmin. This gave support Gaumann's theory that damage of plasma membranes was the primary cause of pathological wilting of plants. Victorin, a host-specific toxin (pathotoxin), produced by *Helminthosporum victoriae* which causes victoria blight of the oats. provided, a lot of information. Permeability damage is a critical event which ensues much before the loss of respiratory control is proved beyond doubt in the experiment with victorin. Respiratory changes occur as the secondary effects of toxin action.

Processes responsible for membrane permeability change due to pathogenesis still not clear. According to one hypothesis these toxins join with some unknown component or receptor in the susceptible cell and thus bring its disorganization. The rupture of the plasma membrane at the site of plasmodesmata has been suggested by Hall and Wood to explain the leakage of electrolyte from endoPGTE treated tissues. Goodman and Klement and Goodman suggested that in the bacterial hypersensitive reaction, an inducer disrupts the tertiary structure of the membrane protein by breaking the disulfide bond. SH-containing compounds, such, as mercaptans, bring about similar splitting. Thus, membrane stability and self-repair may be important determinative factors in susceptibility and resistance.

There are several methods for determining the permeability alterations in diseased tissues. These are:

a. The plasmolytic method which is most commonly used due to its simplicity and non destructive nature.

b. Measuring the electrical conductivity or its reciprocal, resistance of healthy and diseased tissues or of the electrolytes they release, with the help of a Wheatstone bridge.

c. Measuring the potential difference between the interior of the cell and the medium by inserting the tip of an electrode in a cell.

d. Detecting the movement of materials into and" out of the cell by chemical, histochemical or isotopic tracer techniques.

e. Studying the ultrastructural changes in cell walls and membranes by electron microscopy of ultra thin sections.

With the help of above techniques, many diseased tissues of different host pathogen combinations are being studied. The effect of toxin treatment on permeability, specially that of victorin (the site and the nature of effect), is yet to be clarified, and according to Wheeler, "clarification of the mechanism of action of this remarkable substance should shed new light not only on the physiological aspects of pathogenesis but also on the more general problem of the nature of cell permeability."

PROCESS OF PHOTOSYNTHESIS

Photosynthesis is the basic function of green plants that enables them to transform light energy into energy of chemical bonds. Literally, photosynthesis means "synthesis with the help of light". This covers various processes in organic and inorganic chemistry. But the term is usually applied to one reaction only-the synthesis of organic matter by plants in light-a process which is also called carbon assimilation.

In the basic reaction of photosynthesis, carbon dioxide from the air and water from the soil are brought together in the chloroplasts in the presence of light which react to form glucose with the concurrent release of oxygen.

$$6CO_2 + 12H_2O \xrightarrow[\text{Chlorophyll}]{\text{Light}} kC_6H_{12}O_6 + 6O_2 + 6H_2O$$

$$\Delta G° = 686 \text{ kcal.}$$

Carbon dioxide is reduced to the oxidation reduction level of carbohydrate,as symbolized by CH_2O. Each molecule of CO_2 fixed into carbohydrates requires a net input of 1,18,000 calories from light. As an inherent feature of this process, light energy is also converted to chemical energy by a process known. as photosynthetic phosphorylation, where light energy is stored in the useful high energy organic phosphate compound-adenosine triphospbate (ATP).

The details of the change digtorbances in the phenomenon of photosynthesis by plant pathogen have received little attention and there is little information available. The main attention has been paid to the effect of the pathogen on the overall rate of the photosynthe sis of plants because the photosynthetic apparatus-the chloroplast is destroyed or disturbed by pathogens. Our understanding of the disturbances in photosynthesis as a result of infection will not be complete, unless w e study the mechanism in a healthy plant. Our description of the process of photosynthesis in healthy plants will be brief. The pigments, enzymes and cofactors of photosynthesis are all located in chloroplast. The chloroplast, as shown in figure is an ellipsoidal structure.

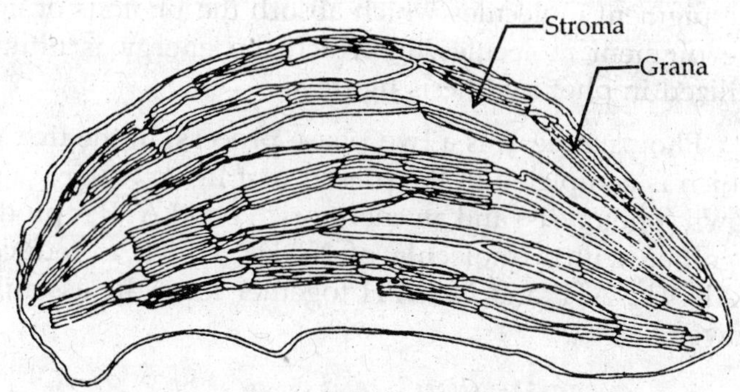

Figure. Submicroscopic structure of chloroplast

It is a highly organized structure in which the pigments are found in a series lamellae in the regions termed grana. A smaller number of lamellae is also found between the grana.

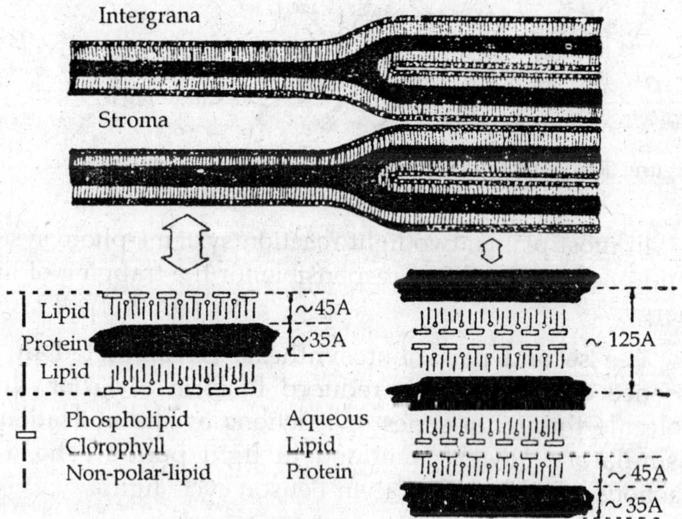

Figure. Enlarged grana lamellae showing position of pigment molecules

A model of lamellae(from corn) depicting its molecular structur is shown in figure.

The lamellae is made up of an ordered array of chlorophll, lipid and of pigment molecules (such as carotene) oriented in such a way as to allow the. transfer of light from the pigment molecules which absorb the photons of light to the pigment molecules, in which the energy is ultimately utilized in photosynthesis figure.

Photosynthesis is a two-phase process. In the first phase which is completed in the presence of light. Water is broken down (photolysis) and an energy-rich bond ATP is produced, together with a molecule of NADP which is reduced to $NADPH_2$. ATP and NADPH together form the assimilatory power.

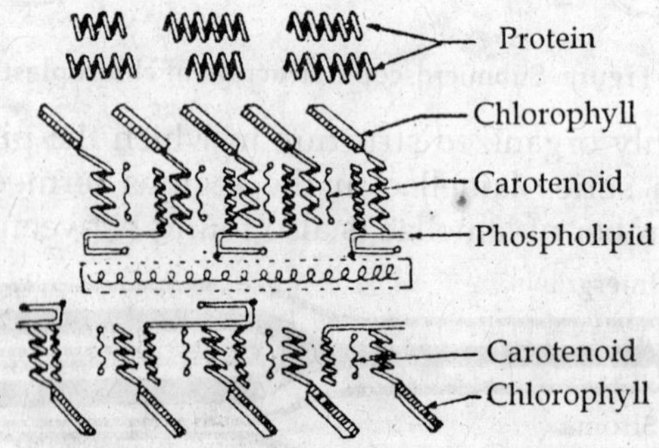

Figure. Representation of molecular arrangement of different pigments, phospholipids and proteins.

In most plants, two light reaction systems-photo system I and photosystem II are responsible for the trapping of light energy.

The second part photosynthesis is completed in the absence of light. CO_2 is reduced to form a carbohydrate molecule through a series of reactions by the utilization of assimilatory power generated in light period. The dark reactions are called the Calvin-Benson cycle figure.

ALTERNATE PATHS OF CARBON FIXATION

The Calvin-Benson cycle of CO_2 fixation into carbohydrates is not the only process by Which plants can fix CO_2. Hatch and Slack discovered that in sugarcane the first stable, intermediate is not a 3-carbon atom compound such as the one which occurs in the Cal in-Benson cycle but a 4-carbon atom compound, oxaloacetate, another intermediate in plant and animal glucose metabolism. The key feature of this C_4 cycle is that the primary carboxyiotion reaction is catalyzed by PEP carboxylase so that the resulting product is oxaloacetate. This 4-carbon acid is formed by a completely different pathway in which a CO_2 molecule is connected with a 3-carbon atom precursor to form the 4-carbon acid.

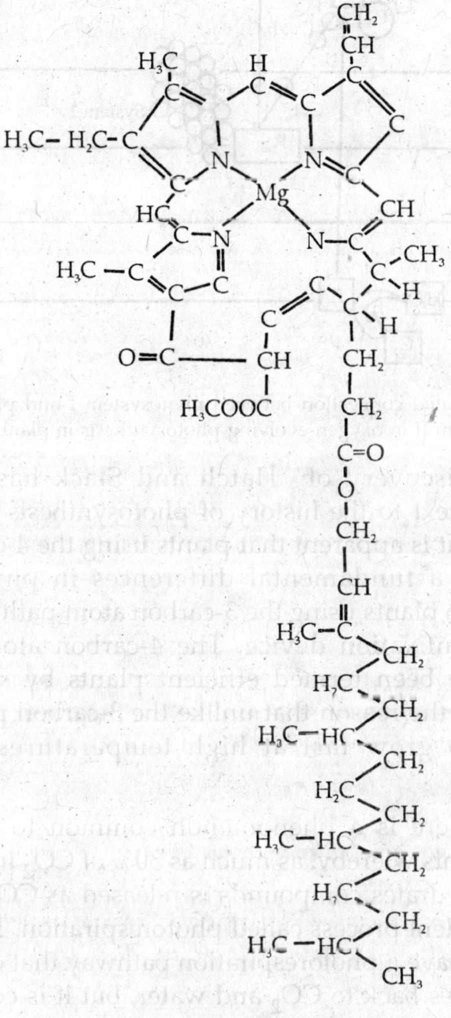

Figure. The chlorophyll molecule

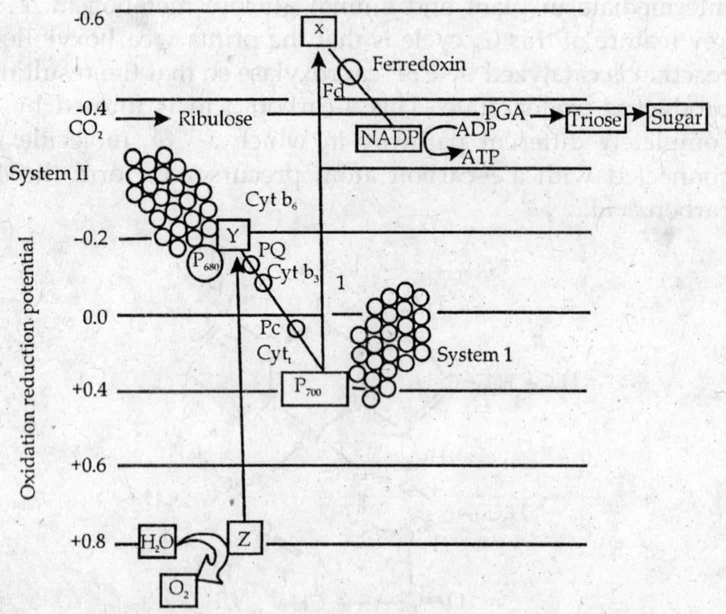

Figure. Postulated connection between photosystem I and photo system II and photosystem II in oxygen-evolving photosynthesis in plants.

This discovery of Hatch and Slack has added an interesting text to the history of photosynthesis research. in as much as it is apparent that plants using the 4-carbon atom cycle have a fundamental differences in physiology as compared to plants using the 3-carbon atom pathway as their main carbonfixation device. The 4-carbon atom pathway plants have been termed efficient plants by some recent workers for the reason that unlike the 3-carbon plants, they are able to grow fast at high temperatures and light intensities.

Also there is a phenomenon common to many non-efficient plants wlereby. as much as 30% of CO_2 initially fixed into carbohydrates compounds is released as CO_2 through a light dependent process called photorespiration. The efficient plants also have a photorespiration pathway that can degrade carbohydrates back to CO_2 and water, but it is compensated for by a very efficient system that traps any CO_2 released by

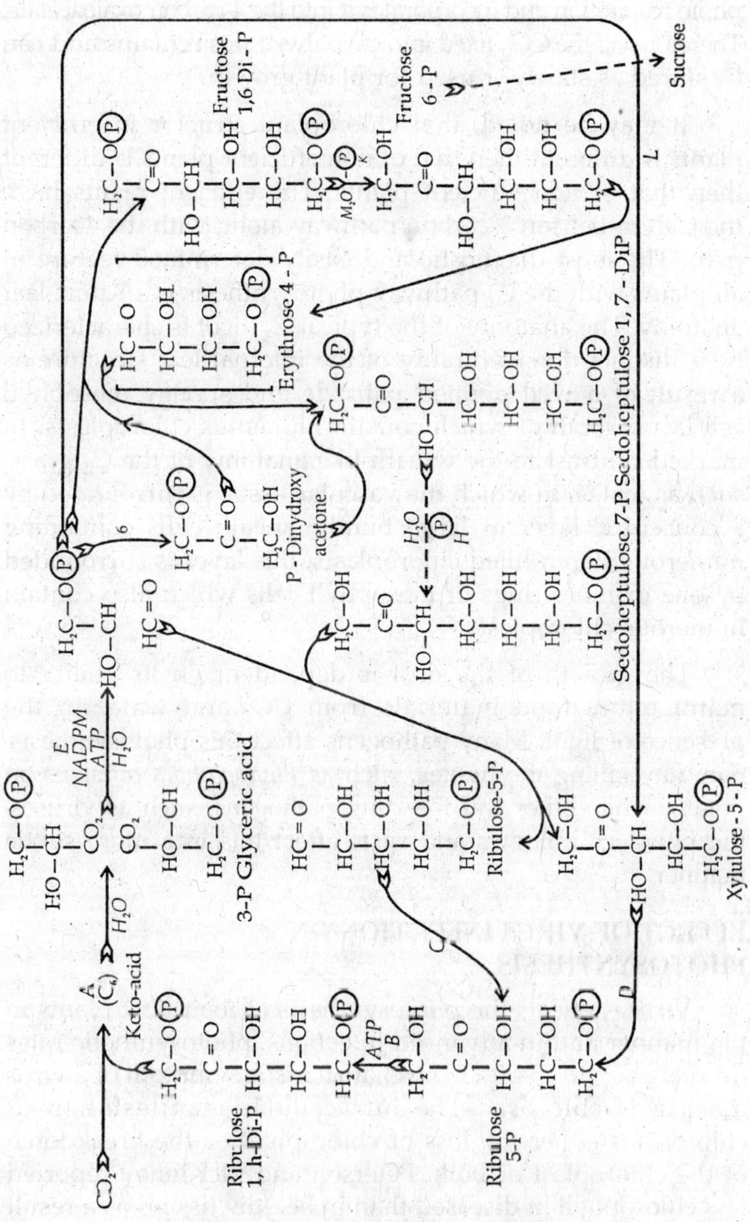

Figure . Calvin-Benson pathway for carbon fixation in green plants

photo respiration and incorporates it into the 4-carbon oxaloacetate. Therefore, all the CO_2 fixed into carbohydrates remains and can be stored as starch or used for plant growth.

It may be noted, that chloroplant structor in efficient plants is different then that of non efficient plants is different then that of non efficient plants. The efficient plants have the Calvin-Benson 3-carbon pathway along with the 4carbon path. The most diagnostic and easily determined feature of all plants with the C_4 pathway photosynthesis is a Kranz leaf anatomy. The anatomy of the typical C_3 dicot is characterized by a distinct dorsiventrality of the internal leaf structure as a result of a well defined palisade and spongy mesophyll cell layers, both of which contain numerous chloroplasts. In marked contrast to the wreath-like anatomy of the C_4 grass, such as maize, in which the vascular tissue is surrounded by a concentric layer of large bundle sheath cells containing numerous starch-filled chloroplasts, this layer is surrounded by one or more rings of mesophyll cells which also contain numerous chloroplasts.

The growth of the host is dependent on its ability to manufacture food materials from CO_2 and water in the presence of light. Many pathogens affect this photosynthesis. Some by killing the tissues, such as *Phytophthora infestans* on potato, while other leaf infecting pathogens, such as viruses, the powdery mildews and rusts affect this in a more subtle manner.

EFFECT OF VIRUS INFECTION ON PHOTOSYNTHESIS

Viruses effects the photosynthesis of their host plants in big manner and in advanced infections, photosynthetic rates decreases up to 75-80%. The characteristic symptom of a virus disease is chlorosis. The intracellular manifestation of chlorosis is either the loss of chlorophyll or the breakdown of the chloroplast or both. Peterson and McKinney reported less chlorophyll in diseased than in healthy tissues as a result of the influence of four mosaic diseases. Roberts *et al.* reported a reduction in photosynthetic rates in the case of those virus diseases in which the symptoms were mild or absent, as well as the induction of chlorosis, mosaic or local lesion formation.

Roberts and Corbett reported reduced photosynthesis in tobacco plants infected with Tobacco Ring Spot Virus. They demonstrated that a symptomless tissue of low virus content showed reduced photosynthesis (evolution) per unit of chlorophyll. This suggests that in this case, the disturbance of photosynthesis does not result from chlorophyll loss; infact, this type of tissue has a slightly increased chlorophyll content per unit leaf area and appears greener than the uninfected tissue. Apparently, other components of photosynthetic apparatus are affected before the conversion of chlorophyll to the chlorine ring containing magnesium (the chlorophyllide) and chain alcohol (the phytol).

$$\text{Chlorophyll} \xrightleftharpoons{\text{Chlorophyllase}} \text{Chlorophyllide+phytol}$$

In Chinese cabbage infected with turnip mosaic virus, virus growth appears to take place in chloroplasts and there is proof that reactions of photosynthesis may be involved in virus synthesis. Goffeau and Bove found increased cyclic and non-cyclic phosphorylation in chloroplasts in Chinese cabbage infected with turnip mosaic virus, and suggested a direct role in viral synthesis forthe ATP produced in this manner.

In certain virus infections, some of the enzymatic processes of photosynthesis are diseaved. The Hill reaction activity, of sugarbeet chloroplast from plants infected with virus yellow and the Hill reaction and photosynthetic phosphorylation in chloroplasts isolated from TMV infected plants are impaired as a result of virus diseases in advanced infections. Enzyme activities of isolated chloroplasts were expressed per unit of chlorophyll and, therefore, a reduction in rate indicates, that more enzyme activity is lost than is chlorophyil from the chloroplast. Bedbrook and Matthews showed that primary fixation reactions of chloroplasts isolated from TMV infected plants see in different from those of healthy plants; 3-phosphoglyceric acid and sugar phosphates were less in quantity and a greater proportion of fixed carbon appeared is phosphoenol pyruvate and C_4 acids.

Duke investigated $^{14}CO_2$ in tobacco leaves infected with TMV and found more label in amino acids, such as serine, alanine and glycine, associated with photosynthesis in diseased than in healthy plants. These amino acids were rapidly attached in to proteins, including viral proteins.

Deranged starch metabolism, with either an excessive accumulation or paucity of starch, is a common biochemical symptom of viral diseases. Starch accumulates in the chloroplasts which are affected by virus diseases. Holmes in observed that tobacco plants infected with TMV failed both to form and to metabolize starch at a normal rate. There was more starch in virus infected areas. This phenomenon was not always consistent; leaves taken at the end of the period of active photosynthesis often contained less starch in virus-infected plants than those taken from the green house and placed in the dark under conditions favourable for the starch to be metabolized. Holmes found that the infected tissues always retained more starch. These observations dearly suggest that TMV infected leaves have a reduced capacity for synthesizing and degrading starch. Robbins found an accumulation of starch in mosaic leaves of sugar beet. Starch accumulation is also characteristic of a number of other viral diseases, for example Potato Leaf Roll, Sugar Beet Yellows and Barley Yellow Dwarf. It is not known where the virus affects these processes, although it can influence the loss of some 'enzyme activity from the chloroplast.

PHOTOSYNTHESIS AS EFFECTED BY BACTERIAL PATHOGENS

An attack by bacterial pathogen frequently involves chlorotic symptoms, for example, a number of leaf spotting xanthomonads where there is a direct effect on chloroplasts. This is perhaps by the toxic metabolites of the pathogen; studies by Braun (1955) on the mode of action of. wildfire-toxin produced by *Pseudomonas tabaci* suggests that the toxin either destroys chlorophyll or inhibits its synthesis.

Photosynthesis may also be affected indirectly by the occlusion of the vascular system by a bacterial pathogen leading to water stress and wilting. This results in the closure of stomata, and consequently, there is inhibition of CO_2 absorption. Backman, Braun and Buddenhagen (1962) ascribed the symptom of wilting in banana induced by *Pseudomonas solanacearum* to one of the two causes: (1) vascular occlusion or (2) a toxic effect which would influence water retention by host cells.

PHOTOSYNTHESIS AS EFFECTED BY FUNGI

Fungal pathogens seem to affect the photosynthetic process by affecting the chloroplast. Macroscopic observations of many fungal diseased tissues also suggest that the chloroplasts are affected. Rusted wheat leaves and leaves of bean infected with *Uromyces phaseoli* retained chlorophyll in regions termed *green islands* at the edge of the infection. Wang (1961) has shown in the case of bean rust, that these regions are photosynthetically active and starch tends to accumulate in them to a greater extent than in the surrounding areas infected by *Uromyces phaseoli,* while Schipper and Mirocha (1969a) showed the depletion of starch in the same infected tissues. Certain effects on the photosynthetic system may be observed anatomically. There may be chlorosis of the whole leaf; this is due to a reduction in the amount of chlorophyll per chloroplast rather than to a reduction in chloroplast number, at least in *Puccinia recondita* on wheat (Manners and Gandy, 1954). Green islands may appear around a pustule when the remainder of the leaf is yellow (Allen, 1942). Bushnell and Allen (1962) found that *Erysiphe graminis* on barley caused-the tissues around a pustule first to become chlorotic and then to redevelop chlorophyll. Green islands might also be induced by extracts of conidia of *E. graminis* or uredospores of *Puccinia graminis.* Auxins or kinetins (Mothes, Engelbrecht and Kulajewa, 1959) may also induce green island formation. Bushnell (1967) showed that in barley kinetin induced islansd contain more starch but less nitrogen than mildew induced islands. Electron micrographs show that grana and intergranar lamellae of chloroplasts in leaves infected with *E. graminis* become progressively more disorganized (Plumb, 1969) with the progress of the disease.

The rate of photosynthesis rises during 24-48 hours following inoculation in wheat I (Doodson *et al.*, 1965) infected by *P. striiformis* and in barley infected by *E. graminis* (Allen, 1942). In the later stages of infection. the rate of photosynthesis fal1s. Doodson *et al.* (1965) found that in wheat plants infected with *P. striiformis,* the rate of fall is greater than the rate of decline in chlorophyll content, that is, the assimilation rateunit of chlorophyll drops. This was similar in the case of barley infected with *E. graminis.* Wynn (1963) showed that photosynthetic phosphorylation per unit of chlorophyl was not affected, indicating that subsequent to phosphorylation some process must be damaged, In the beginning there is a slight initial stimulation of the photosynthesis of inoculated leaves but this is soon followed by a rapid decline (Figure).

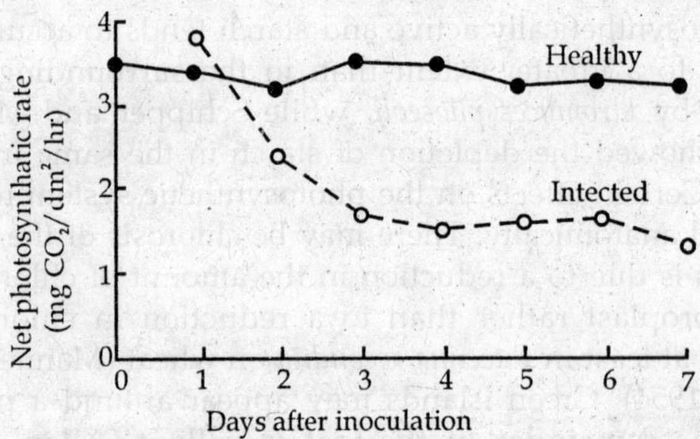

Figure. Changes in photosynthesis of oak leaves following infection by the powdery mildew fungus Mycosphaerella alphitoides. (From C.H. Dickinson and J.A. Lucas, Plant Pathology and Plant Pathogens. Blackwell Scientific Publications, Osney Mead, Oxford, 1977, p. 94)

According to Magyarosy *It ill.* (1976), the rate of 1402 uptake by leaf discs from sugar beet infected with powdery mildew is also reduced as compared to healthy tisues (figure).

Livne (1964) found inhibition of photosynthesis at 0.06 %CO_2 concentration in the primary leaves of bean infected with *Uromyces phascoli* var. *typica* in the late stages of infection. Initial inhibition was found to be dependent on the intensity of infection. When the infections were heavy (20-300 pustules per cm) photosynthetic rates in the infected and healthy primary leaves were the same till the fifth day, by which time the first symptoms appeared. When the infecion sites were comparatively less (lflO to 1/3), there was slight stimulation in photosynthetic rates and this continued till sporulation began. In plants inoculated in such a way as to deposit inoculum only on tpe primary unifoliate leaf in cases of heavy infections, a stimulation of 50 to 100 % over control was observed in uninfected trifoliate leaves of the plant while only a very slight stimulation was noticed in the same healthy leave of the, infected plants (Figure).

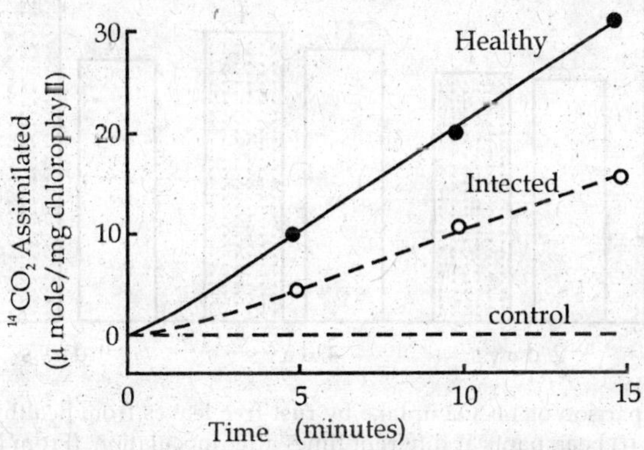

Figure. Effect of infection by *Erysiphe polygon* on the rate of photosynthetic $^{14}CO_2$ assimilation by sugar beet leaf discs. (From Dickinson and Lucas, *Plant Pathology and Plant Pathogens,* Blackwell Scientific Publications, Osney Mead, Oxford, 1977, p. 95.)

Doodson *et a1.* (1965) obtained results in rusted wheat leaves similar to those of Livne for lightly infected bean leaves.

A connected effect of infection intensity was seen for *Aibugo candida* on *Brassica juncea* cotyledons. In think infections, Black found diminished rates of photosynthesis, which started four days after infection and before the starting of sporulation. Harding reported that in the case of low intensity infections the rate of photosynthesis was unchanged for at least seven days after infection.

It is tough to determine the process responsible for

changes in photosynthetic rates. The inhibitions in photosynthetic rates in the late stages of rust and mildew infections might be explained by the degradation of the photosynthetic apparatus. Thus, damage to the photosynthetic apparatus is not responsible for the low rates of photosynthesis in diseased tissues at normal CO_2 levels. The biochemical reactions responsible for either stimulations or inhibitions are also vague.

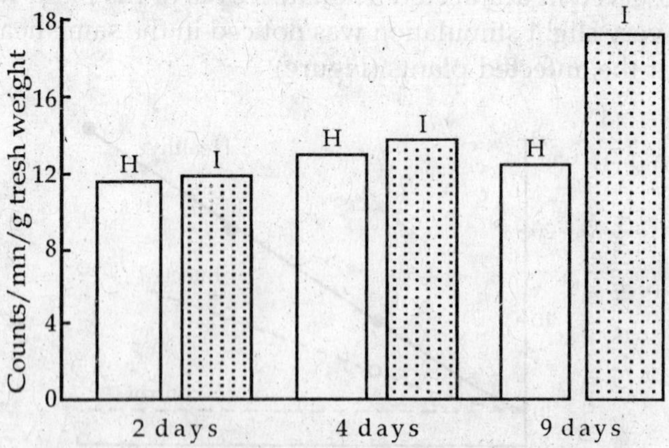

Figure. Comparison of 14CO2 uptake by rust-free leaves from healthy (H) and rust-infected (I) bean pants at different times after inoculation. (From Dickinson and Lucas, Plant Pathology and Plant Pathogens, Blackwell Scientific Publications, Osney Mead, Oxford, 1977, p. 95.)

Scott and Smillie (1966) reported a decrease in the activity of some enzyes involved in photosynthesis but since the same enzymes are associated with respiration (*i.e.* aldolase) their role in photosynthesis decline is not definite.

In the case of facultative parasites, recent studies have shown a significant decrease of photosynthesis within a day after the inoculation of wheat with *Septoria nodorum* and of ragi with *Helminthosporium nodulosum*. This supprering effect suggests that there are other factors involved in the decrease of photosynthesis other than chlorophyll reduction. In *Verticillium* wilt of cotton, Mathre reported a 50 o reduction of photosynthesis for apparently healthy leaves with normal chlorophyll content.

Changes in starch content following infection have been obsened in many foliar diseases. The general pattern is decline in beganning followed by a marked increase with heavy, accumulations around the margins of lesions. Still later, the starch content again decrease. Starch accumulates in wheat leaves infected with stripe rust. Work with wheat leaves infected with *Puccinia striiformis* led to the suggestion that changes in the activity of ADPglucose photophosphorylase during disease development might account for the pattern of changes in starch contents. MacDonald and Strobel assayed these enzymes in the presence of activators (sugar phosphates) and inhibitors (inorganic phosphates) at concentrations found in diseasd leaves at diference times after infection. They suggested that the method of changes in the activity of this enzyme was similar to that of changes in starch contents. Changes in starch in leaves infected with *P. striiformis* were very alike to those observed in other diseases. In contrast, photosynthesis in infected leaves increases sharply, reaches a maximum of twice the normal as symptoms appear, and remains above normal for, at least six days after sporulation. These results which differed markedly from those obtained from other diseases, indicate that *P. striiformis* infected/plants may represent a special case.

The facultative parasites, such as *Cercospora,* also induce a depletion of starch. Kolandaisamy reported the depletion in starch content in banana leaves and an initial go in major content of the ground nut leaves infected by C. *personata.* However, there was a sharp decline in sugar content. Although increased respiration seen in the tissues infected with facultative parasites may be a factor in the depletion of sugars, it is also possible that, as pointed out by Gaumann the sugars may form the basic materials for the growth and sporulation of the pathogen.

Plants infected with rusts or powdery mildews fix the greatest amount of CO_2 in the dark at the time ,of sporulation. This was shown by exposing the leaves to $^{14}C0_2$. Bean and cereal rusts uredospores have the malic enzyme which catalyzes the reversible reaction, pyruvate + $CO_2 \leftrightarrow$ malate, in both directions. According to Mirocha and Rick this enzyme may be responsible, at least in part, for increased dark CO_2 fixation in diseased leaves.

The proof for increased dark CO_2 fixation in diseased plant tissues in the absence-of a living pathogen was provided by Luke and Freeman by measuring the Krebs cycle intermediates in victorin treated oat leaves held in the dark in the presence and absence of CO_2, Luke and Freeman suggested that cations released from the vacuole as the result of victroin induced disruption of permeability might increas the CO_2 fixation.

Mirocha has reviewed the effects of infection or treatment with pathotoxins on dark CO_2 fixation. the HC-toxin produced by *Helmilthosporium carbonum* applied to maize (C4 plant) causes increases indark CO_2 fixation similar to those obtained when oats (C_2 plants) are treated with victorin According to Mirocha both toxins may act by increasing the turnover of Ribulose-5-phosphate and Ribulose-1-5-diphosphates, thereby increasing the rates of CO_2 fixation.

RESPIRATION UNDER INFLUENCE OF PATHOGENS

Respiration is the inderpeusabli condition for life. In normal plant cells the function of respiration is twofold-to provide (a) energy and (b) carbon skeletons. Respiration is the oxidationi-caerobic or anaerobic breakdown of organic material, to simple compounds, H_2O and CO_2 with the production of utilizable energy for the cell in the form of ATP. In aerobic respiration this takes place vith the uptake of oxygen from the air. In fermentation the degradation of organic material is anaerobic. Energy is given through the process of respiration. Actually, the oxidation of carbohydrates and related compounds is the main source of energy, but energy derived from photosynthetic phosphorylation is also a feature of plant metabolism. The energy rich group generated by oxidation is capable of starting reactions requiring energy. So the bonding of the chemically bound energy to other systems is the key reaction in the hiosynthetic processes of plants.

One of the simplest substrate for oxidation is glucose. The total reaction of respiration of glucose is as follows:

$$C_6H_{12}O_6 + 6O_2 = 6CO_2 - 6H_2O + \Delta G° = 686 \text{ kcal.}$$

This respiratory process, can be divided into two parts. These are

a. The initial phase known as the Embden-Meyerhof-Paranas pathway or glycolysis, in which each glucose molecule is divided into two molecules of pyruvic acids.

b. The end phase (citric acid cycle or Krebs cycle or Tricarboxyic acid cycle) in which pyruvate is oxidized to CO_2 + H_2O, using oxygen as the final hydrogen acceptor.

The glycolytic phase can take place in the absenct of O_2 (anaerobically) and is common both to aerobic respiration and to alcoholic or lactic acid fermentation end phase is totally aerobic.

If enough amount of oxygen not present and if anaerobic pathways are favoured, this scheme is changed. Glycolysis occurs but intermediates of glycolysis acting as the final hydrogen acceptors are reduced to other end products. A typical overall reaction for anaerobic respiration is:

$$C_6H_{12}O_6 = \underset{\text{(ethyl alcohol)}}{2CH_3CH_2OH} + 2C0_2 + 2 \text{ ATP}$$

Note that the energy yield of fermentation is less because the end product is incompletely oxidized, and still stores energy. The discussion of respiration is divided into (a) glycolysis, (b) fermentation end products, and (c) CA cycle.

PROCESS OF GLYCOLYSIS

Glycolysis is the process of conversion of glucose to pyruvic acid by any biochemical pathway. The most common glycolytic pathway is the Embden-Meyerhof Paranas pathway, although the Hexose monophosphate pathway

(HMP) and the Entner Doudoroff (ED) pathway may supply alternate mechanisms. The biochemical differences of these pathways is given in Figure.

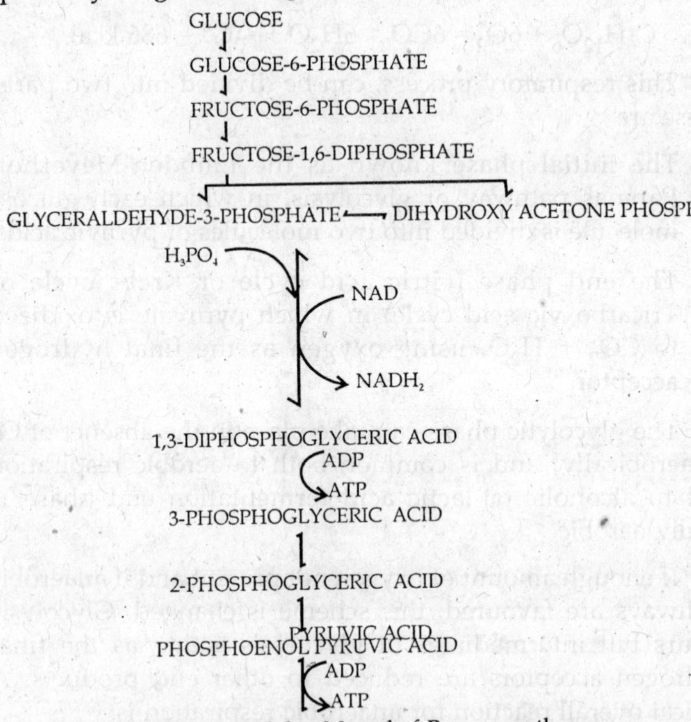

Figure. Embden-Meyerhof-Paranas pathway.

The interconnected relationships between the EM, HMP and ED glycolytic pathways is given in Figure. The EMP and HMP pathways occur in animals, plants, bacteria: and fungi, while the ED pathway is restricted to bacteria and fungi.

The EMP pathway in fungi is the principal glycolytic pathway as in most of the other organisms and usually accounts for at least 50% of the glucose break down, in some fungi it accounts for all the glucose break down. However, the EMP pathway may be secondary in importance to either the HMP or ED pathway in a few fungi or totally lacking, as in the case of *Cladariornyces fumago.* The EMP pathway becomes increasingly important under anaerobic conditions as the HMP pathway accounts for all the glucose dissimilated.

Ths HMP pathway is common in plants and animals.

The enzymes needed to carry. out the HMP pathway are widely

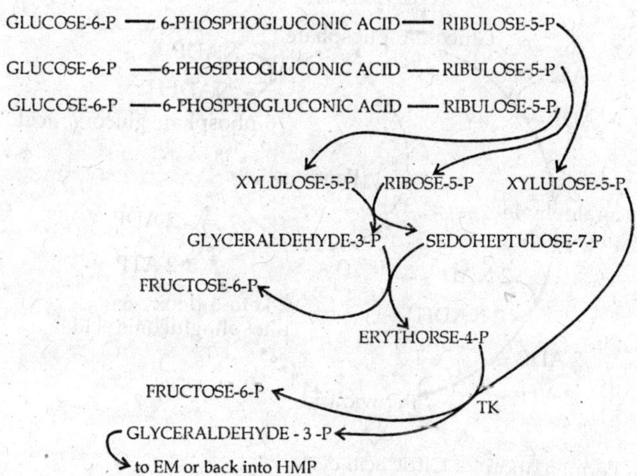

Figure The HMP pathway

present in fungi and may be universal. The presence of the enzymes, however, is not a proof that the HMP pathway occurs. It is found in a number of fungi, usually accounting for less than 40% of the glucose break down. The relatively low level of the H M P reactions is perhaps due to limited amounts of NADP. The HMP pathway is the major route of glycolysis in 5 of the 15 fungi surveyed.

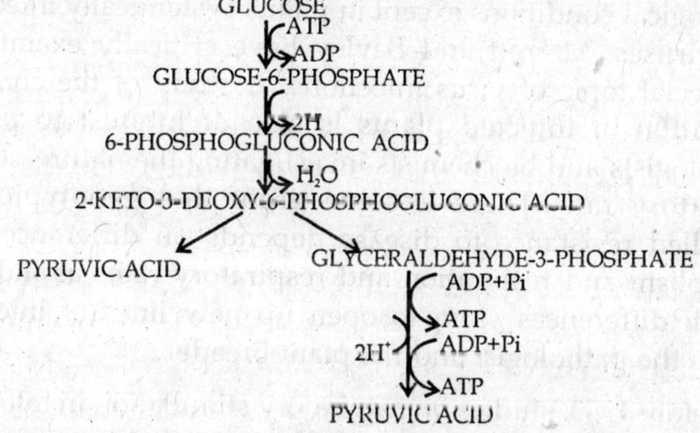

Figure. The Entner-Doudoroff pathway.

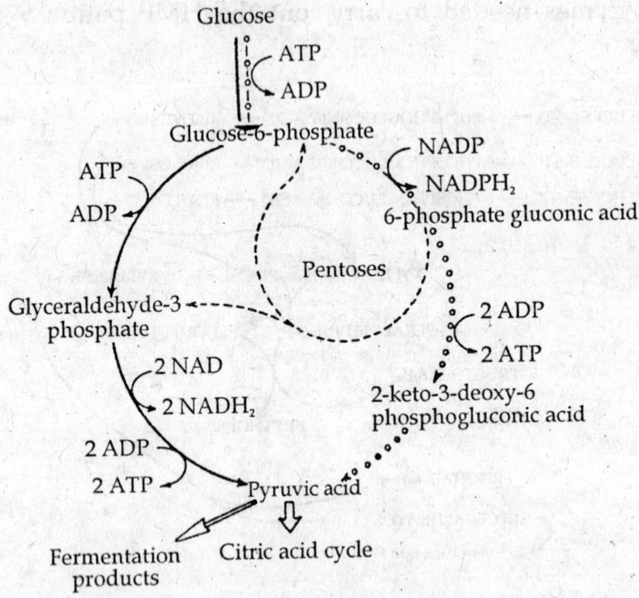

Figure. The interrelationships between EM, HMP and ED glycolytic pathways.

RESPIRATION IN PLANTS UNDER INFLUENCE OF PATHOGENS

Plant respiration as an indicator of disease has been a under investigation for a number of years. Increased respiration in tissues generally accompanies a variety of pathological conditions except in plants systemically infected with viruses. Merrett and Bayley have critically examined the special topic of virus infections. A study of the change respiration of infected plants is also of interest to plant physiologists and biochemists investigating the nature of the respiratory rate. There is some proof that genotypically controlled resistance to disease depends on differences in metabolism and respiration and respiratory rates as indices of such differences seem to open up new lines of interest both to the pathologist and the plant breeder.

Before 1953, studies on respiratory stimulation in infected plants were limited generally to the quantitative aspect of the increase. It was Allen who first studied that disease was the result of disturbances in host metabolism induced by the toxins of the parasite. He claimed that increased respiration in mildew disease was the result of a toxin produced by the fungus which uncouples phosphorylation from electron transfer in host cells. Under given figure shows the rate of oxygen intake in cabbage cotyledons during colonization by *Peronospora parasitica.*

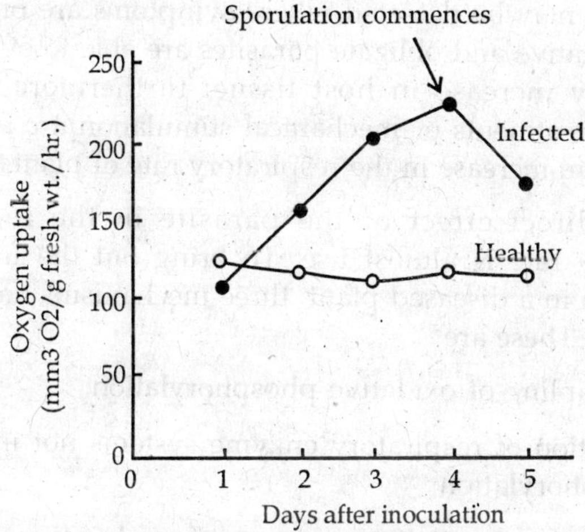

Figure. Oxygen uptake by healthy cabbage cotyledons and those infected by *Peronospora.*

INCREASE MECHANISM OF RESPIRATORY

Increase in respiration is not a phenomenon singly associated with a few plants damage of by specific pathogens but a general responsive reaction of the plant tissue attacked by pathogenic micro-organisms, including fungi, bacteria and viruses.

A large number of diseases are characrerized by changes in respiration. It has been observed in potatoes infected with *Penicillium* spp., *Phytophthora infestans;* sweet potatoes attacked by *Rhizopus tritici, Ceratostomella fimbriata, Gibberella saubinetti* or *Helicobasidium mompa,* A similar respons has also been observed in rice and wheat plants infected by several fungi. Perhaps the most, widely recorded interference in plant tissues is in virus infections because viruses are totally dependent upon the host for their multiplication. The respiration rate of Plants is generally increased by infection, sometimes within hours of inoculation. but large increases (upto 50%) typically occur shortly before the symptoms appear. The amount by which respiration increases is largest for viruses or virus strains that cause serious symptoms and no change may be detected where symptoms are only light both facultative and obligate parasites are able to evoke this respiratory increase in host tissue; furthermore, simple chemical treatments or mechanical stimulation are also able to induce an increase in the respiratory rate of plants.

The direct effect of the parasite in the increased respiratory rate is almost use. To bring out the increased respiration in a diseased plant, three mechanisms have been suggested, These are

a. Uncoupling of oxidative phosphorylation.

b. Induction of respiratory enzyme systems not involving phosphorylation.

c. Augmented synthetic process (acceleration of ATP utilizing systems).

UNCOUPLING OF OXIDATIVE PHOSPHORYLATION

The hypothesis that the pathological increases in given respiration result from an uncoupling of phosphorylation from electron transport was given by Allen on the basis of results obtained earlier by Sempio. The data obtained by Sempio showed that ratios of anaerobic to aerobic carbon dioxide production were markedly lower in diseased than in healthy tissues. Allen interpreted this to mean an abolition of an apparent Pasteur effect in healthy tissues. Allen showed that the respiratory increase of rusted wheat leaves could be due to the action of a phytopathogenic toxin which uncouples the oxidative phosphorylation of the host tissue. It should be noted that ratios of anaerobic to aerobic carbon dioxide production greater than 0.33 but. below 1.0 is merely an indication that the aerobic rate of glycolysis has been dampened by a metabolic control, a situation first described by Pasteur.

Hard evidence for the Pasteur effect requires that such conservation be established by appropriate chemical analysis. Rusted wheat leaves failed to respond to an uncoupling agent 2,4-dinitrophenol or DNP as would have been the case if uncoupling had already occurred. These infected leaves were also insensitive to malonate (a competitive inhibitor of succinic dehydrogenase, thus preventing the conversion. of succinate to fumarate) but remained as sensitive to fluoride (an inhibitor of metal containing oxidases which inhibits the reaction in the EMP pathway) as did healthy leaves Homogenates of diseased leaves also showed a big increase in ascorbic acid oxidation.

Millerd and Scott gave data showing that an uncoupling factor could be isolated from barley infected with powdery mildew and was responsible for the increase in respiration of barley leaves infected with *Erysiphe graminis* var. *hordei.* The results obtained by Samborski and Shaw and Shaw and Samborski were also interpreted as a host mediated respiratory rise in wheat infected by *Puecinia graminis tritici.* The uncoupling of oxidative phosphorylation has been reported to occur in necrotic lesions produced by TMV in *Nicotiana* leaves.

The results from several laboratories failed to confirm reported increases in inorganic phosphate coupled with decreases in organic phosphate in diseased tissue. Heitefuss and Fuchs found more organic phosphate in infected susceptible leaves. Moreover, the incorporation of P^{32} into organic phosphores compounds coincided with an increase in respiration.

Increased synthetic activity an in some cases, increases in the growth of diseases due to by biotrophic fungi give the most cliching proof against the uncoupling hypothesis.

Miller and Koeppe gave evidence that uncoupling plays a role in the leakage of ions caused in Texas male sterile corn inbreds by the selective toxin of *Helminthosporium maydis, race T.* Culture filtrates of the pathogen caused the swelling of isolated mitochondria and an increased oxidation of malate with the loss of respiratory control. The tissue damage inhibition of root growth, respiratory increase and ion leakage can be explained by the effects of race T-toxin on the mitochondria of susceptible plants. Yet the ultimate role of mitochondria in the diseases *in vivo* remains uncertain.

SYNTHETIC PROCESSES

Another reason for the increase in the mechanism of respiration may be due to active metabolism accebraty cups the A.T.P using host tissues. The synthetic processes requiring energy, accelerate ATP breakdown to ADP which in turn causes, respiration to rise. In support of the above mechanism, many characteristic features indicating the activation of the metabolism of the host have been reported, such as:

a. Accumulation, mobilization and synthesis of phosphorus and carbon compounds.

b. The growth of the host tissue.

c. Synthesis of proteins, including activation of enzymes and enzyme systems.

d. Increase in protoplasmic work, such as protoplasmic streamirig (cyclosis).

According to the results of Daly and Sayre. Safflower infected with *Puccinia carthami* showed a pathologically stimulated cellular growth and a concomitant increase in tissue respiration. The rise was at peak during the sporulacion of the parasite. In supporting this abnormal growth of diseased host cells, ATP is consumed; thus the breakdown of ATP is increased and eventually increased the respiratory rate of the host.

Stimulated growth, accumulation and mobilization of metabolites occur in rust and bean plants synthesis of proteins in rotted sweet potato and late blight of potato and increased protoplasmic streaming in late blight of potato. All these demand energy, that is ATP and thus respiration is automatically controlled through the control of the level of ADP (phosphate acceptor).

Besides the respiratory rate, the temperature of diseased tissues is also higher than that of healthy ones. The increase in temperature is from 0.3 to 0.7°C for mildew and rust diseases (Yarwood, 1953). It is not yet known whether the respiration of infected plants is less efficient than that of healthy plants. It is possible that the high rate of respiration of infected plants is, in part at least, a wasteful one.

Eradication of the Pasteur Effect

The uncoupling of phosphorylation or an increase in the rate of synthetic reaction needing energy, raises the level of ADP and inorganic phosphorus in tissues and results in an increase in the rate of respiration. Either of these processes is also responsibie for the abolition of the Pasteur effect. DNP, which acts as an uncoupler of oxidative phosphorylation, also abolishes the Pasteur effect. It was Sempio who emphasized the importance of this phenomen on. Pasteur effect is defined as the inhibition of fermentation in living organisms under aerobic conditions, The abolition of this effect means poor utilization of carbohydrates even in the presence of oxygen. In effect, this amounts to aerobic fermentation. This the respiratory efficiecy of diseased tissue decrease due to poor utilization of glucose An important gain to the plant by the abolition of the Pasteur effect is the accumulation of the intermediates of carbohydrates breakdown which are utilized in the formation or phclic compounds. Phenolic synthesis is triggered in the diseased host tissue.

ALTERATION IN RESPIRATORY PATHWAYS DURING INFECTIONS

The EMP pathway is the most crucial pathway of carbohydrate break down but this respiratory pathway becomes changed in higher plant tissues as a result of biochemical transformations during disfferentiation and maturation, and it is not surprising that infection with viruses, bacteria an fungi also bring about such changes. The HMP pathway is not of major importance, except in special conditions. The work done in this sphere on healthy and infected tissues strongly suggests that the HMP pathway becomes quite important during infections.

Daly have put forward three arguments for a change in the nature of respiratory pathways in rust diseases.

a. It was seen that at sporulation the ratio of anaerobic to aerobic carbon dioxide production was below the limit of 0.33 which was expected if only glycolysis and TCA cycle were operating.

b. Sodium fluoride inhibits the reaction in the EMP pathway by which phospheonol pyruvate is formed from phosphoglycerate through the activity of enolase. Inaction to sodium fluoride showed that the EMP pathway is not operative.

The tracer methodology has been used for this purpose by which radioactive carbon is incorporated into either the C_6 or the C_1 position of the glucose molecule. The ratio of labelled CO_2 released from each source during respiration is measured which indicates the predominating pathway. Activation of the pentose phosphate pathway leads to an increased contribution from the C_1 position, and hence, the C_6/C_1 ratio is lower. In infected plants, the C_6/C_1 ratio is lower than the value obtained from healthy tissues, suggesting that there is an increased participation of the pentose route.

A shift from the glycolytic to HMP pathway has been detected in several diseases caused by fungi, bacteria and viruses (only in local lesion type). Some of the examples are as follows:

a. Infection of *Datura stramonium* by TMV.

b. Tomato tumour caused by *Agrobacterium tumefaciens*.

c. Halo blight of tobacco caused by *Pseudomonas tabaci*.

d. Rusted and mildewed wheat.

e. Potato tubers infected with *Phytophthora infestans.*

Significant decreases in C_6/C_1 ratios following infection were reported by Shaw.

Inhanced activity of the pentose phosphate pathway is suggested for bean hypocotyl tissue infected with *Rhizoctonia solani.* Rawn and Wheeler demonstrated that the *C6/C* 1 ratio of infected oats dropped from approximately 1.0 to 0.6 at the end of four hours of treatment with pathotoxin victorin However, infection of cut sweet potato tuber results in an increase in C_6/C_1 ratio (Akazawa and Uritani, 1962). In the beganning of infection, the ratio is equal both in cut and diseased tissues, but 72 hours after inoculation there is a marked increase in the ratio in infected tissues which shows an increased operation of the glycolytic pathway.

The actual significance of the switch to the pentose pathway is linked to its role in the biosynthesis of different compounds. In addititon to providing pentoses for the biosynthesis of nucleic acid, pentose phosphate intermediates are involved in the production of numerous aromatic compounds, particularly phenols and their derivatives.

Changes in the operation of the Krebs cycle have been seen in several plants by studying the effect of malonic acid which, inhibits an enzyme system (succinic dehydrogenase) in the cycle on oxygen uptake. Some of the examples are as follows:

1. Rusted and mildewed wheat.
2. Potato tuber infected with *Gibberella saubinetii.*
3. In *Cercospora* infection on ground nut. particularly in later stages of infection.

ALTERATION IN CARBOHYDRATES AND ENZYMES

A failure to detect difference in the carbohydrates in the beganning studies was primarily due to the fact that in the chomatographic analysis of carbohydrates, special precautions are to be taken because mannitol (a pnentitol) has Rf value similar to that of sedoheptulose. Daly (1962) showed that glucose labelled in C_2 or C_6 positions indicated that the carbohydrate arabitol was derived from the pentose pathway and the amounts stored in uredospores can account for the drop in the C_6/C_1 ratio of infected tissue.

Attempts have also been made to defect alternative pathways of respiration through changes in appropriate enzymes. In the case of biotrophic organisms, infection results in much higher amounts of the first two enzymes of the oxidative pentose pathway, glucose6-phosphate dehydrogenase (C_6P) and 6-phosphogluconate (6PG) dehydrogenase, but exclusive studies of the other enzymes of the pathway, such as transketolase, are lacking. Increased activity has also been observed in the enzymes of glycolysis (such as aldolase and hexokinase) during infection.

TERMINAL OXIDASES AND ELECTRON TRANSPORT

Cytochrome, polyphenol, ascorbic acids and glycolic acid oxidases are the four main oxidases in most plant tissues which can account for electron transport in the respiration of normal plants.

Cytochrome Oxidase

This is an important terminal oxidase of the cytochrome system which is based in the mitochondria. Increased cytochrome activity parallels increased respiration coupled with oxidative phosphorylation.

Ascorbic Acid Oxidase

Ascorbic acid oxidase catalyses the oxidation of ascorbic acid with molecular oxygen. Its physiological role in plants is little known and its involvement in electron transport is uncertain. In infected plants, ascorbic acid oxidase has been studied only to a limited extent. Farkas and Kiraly and Goodman have reviewed the previous work. An increase in ascorbic acid oxidase has been observed in cereals infected by obligate parasites.

Glycolic Acid Oxidase

Glycolic acid is able to transfer hydrogen to molecular

oxygen and to other acceptors. Glycolic acid oxidase is found only in the green parts of plants and is based in the peroxisomes. Glycolic acid is produced in chloroplasts during photosynthesis. Glycolic acid oxidase plays a prominent role in the peroxisomes in the transformation of photosynthetically, produced glycolate and may also be involved in photorespiration. Glycolic acid is oxidized to glyoxylate in the peroxisomes during photorespiration.

As a result of a decline in the activity of glycolic acid oxidase, photorespiration also declines and this in turn increases the efficiency of photosynthetic CO_2 assimilation.

Polyphenoloxidase Activity

Generally, phenol oxidases catalyze the oxidation of phenolic substances with molecular oxygen. A considerable inhancement in polyphenoloxidase activity has been seen in the stem tissues of *Fusarium* infected tomato plants. Kiraly has found polyphenol oxidase activity in leaf tissues of wheat varieties resistant to *Puccinia graminis* while no such activity was seen in the susceptible variety. The increased phenolase activity in diseased or wounded areas of plant tissues is generally accompanied by an increased concentration of phenolic substances in diseased plants. Polyphenol oxidase activity is considered to be involved in the hypersensitive reaction of resistant varieties. The oxidation products of phenols are toxic to the host cell as well as to the pathogens.

RELATIONSHIP OF RESISTANCE AND SUSCEPTIBILITY

Many scienfists have studied the relationship between respiration and resistance to plant pathogenic organisms. As early as 1912, Bach suggested that the oxidative enzymes of the host might protect infected tissues by oxidizing the toxic substances of parasitic origin.

From his studies of the mildew of wheat, Sempio developed the concept of "metabolic resistance" to parasitic attack. According to him, it is the ratio between synthetic reactions like photosynthesis, and breakdown reactions like glycolysis and respiration that decide the outcome of infection, the method of respiratory rate and changes with the advancement of the disease. Samborski and Shaw observed that in wheat rust, the resistant variety recorded a sharp rise followed by an equally sharp fall in respiration, while the susceptible variety had a continuous rise in respiration.

Some interesting correlations between the respiratory rate and resistance to parasitic attack have been observed at Madras laboratory. Andal studied the respiratory rate of three varieties of rice differing in their of intensity susceptibility to *Fusarium moniliforme.* It was found that the respiratory rates of both root and shoot tissues were directly proportional to the intensity of susceptibility. Similar correlations were seen by Lakshmanan in cotton varieties resistant and susceptible to *Fusarium oxysporum* f. sp. *vasin fectum.*

Comparisons between the tuber tissues of resistant and susceptible plants appear to show that incompatible interactions showed an earlier and sharper rise in respiration. The situation between resistant and susceptible wheat infected with rust is same.

Swamy studied that in the case *of* groundnut plants with different degrees of susceptibility to *Cercospora personata,* the resistant variety TMV3 showed a higher rate *of* respiration, than did the susceptible variety TMV2. Barley leaves infected by powdery mildew exhiting a similar case. In the roots of the susceptible cotton variety infected with *Fusarium vasinfectum,* Lakshmanan studied that a anhancies in respiration occurs only from the fourteenth day after inoculation. In the roots of the resistant varieties and susceptible varieties rendered resistant by the addition of zinc to the soil, even on the seventh day, infected plants show a higher rate of respiration than healthy ones and the rates fall thereafter and approach that of the controls. In the the stem tissues, the situation is essentially similar to that in roots.

When dead local lesions are produced by TMV in *Nicotiana* leaves, the enhancement seems to occur mainly from oxidative phosphorylation. When dead local lesions form in inoculated *Nicotiana* leaves, the upper uninoculated leaves become more resistant to infection and the lesions produced by TMV are few and small than those induced in previously virus-free plants. The development of this resistance is correlated with increased peroxidase activity in uninoculated leaves. It is believed that the increased peroxidase activity increases the rate at which quinones can be produced and cause necrosis at infection points, and hence, the spread of the virus is stopped sooner.

ALTERNATION IN NITROGEN METABOLISM OF INFECTED PLANTS

In many diseased tissues, especially in those which shows characterstics, like, overgrowths, galls or other growth abnormalities, increased metabolic activity is accompanied by increases in nucleic acid and protein synthesis. in considerable amount Cytochemical evidence for major changes in nuclei and nucleoli and changes in ribosomes is available.

Nucleic Acid Metabolism

No field of biological science has developed and progressed to such an extent during the last few decades as have nucleic acids because they are involved not only in the transfer of genetic information but also in the regulation of gene activity. Changes in the nuceic acid level have been seen in diseases caused by every type of pathogens. The host RNA content in TMV infected tobacco leaves increases shortly but afterwards the viral RNA increases together with the total of RNA. Total RNA increases in infected tissues, while it diminishes in the uninoculated control tissues. In various virus-host. plant combinations, transitory increases, moderate declines; or no significant changes in DNA have been reported. Such variations probably reflect differences among the plants and viruses studied. In the case of tobacco leaves infected with TMV, one scientist or reported that t-RNA increased fourtimes after infection.

Gall cells produced by various fungi, bacteria and nematodes have been shown to contain a vart quantity of nucleic acids. In the case of *Plasmodiophora brassicae,* causing club root disease of crucifers, the infected cells each contain a big nucleolus which may be as much as 30-or more fold in volume over nucleoli in non-parasitized cells and the DNA content may be sixteeen times greater than in the non-infected cells. Hypertrophy of host cells together with enhancement in the sizes of nuclei have been seen in various other host parasite combinations. In the same disease, the nucleous of root hair cells begins to enlarge within two hours after penetration and after twenty hours this volume has increased sixtimes. The increase in nucleolar volume has accompanied by an increased nuclear RNA,and non-histone protein. The lysine rich portion of histone decreases. Since histones have a regulatory role, any change in their quantity is related to a change in the regulation of gene action. It seems that the normal transcriptional processes of invaded cells are changed by parasites.

The higher level of RNA in rust infected wheat is shown also by an enhanced incorporation of p32 into RNA during the first four to five days after inoculation (Figure). An increase in the RNA content of the nuclei of the cells of

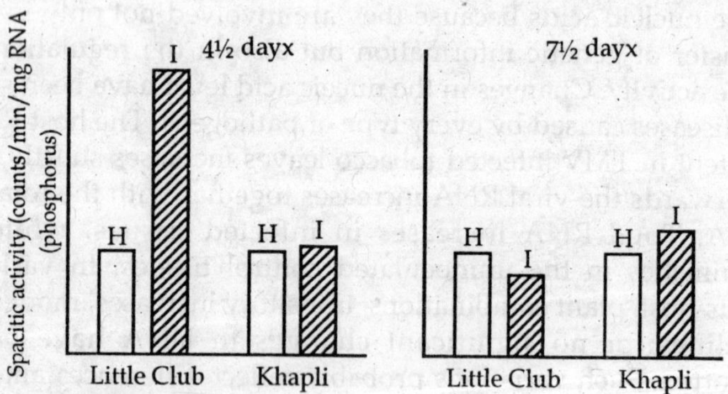

Figure. Amount of P22 incorporated into RNA of wheat leaves CV LIttle Club (compatible) and CV Khapli (incompatible) inoculated *with Puccinia graminis.* H=healthy and I=infected leaves.)From Dickinson and Lucas, Plant Pathology and Plant Pathogens, Blackwell Scientific Publications, Osney, Mead, Oxford, 1977, p. 109)

diseased plants probably indicates an induced synthesis of

proteins in the host. There is in beganning swelling and later breakdown of the infected host nuclei in many fungal diseases of the plants This is associated in every case by an increase in RNA and protein synthesis. Eventually the collapse of the nuclei is due to by proteolysis in the host. A decrease in the RNA and DNA contents of the soyabean leaves infected with *Peronospora manschu rica* has been demonstrated by Millikan *et al.* Callow showed that an increase in the total RNA content of cucumber leaves during the period of leaf expansion is retarded by inoculation with *Erysiphe chicoracearum*. But when the expansion of the leaves went up to 75% of their final area, a very slight increase in cytoplasmic RNA was observed.

Alteration in Protein Metabolism

Changes in protein metabolism in plant tissue during the beganning of pathogenic infection and course of disease development have been reviewed by Uritani. The total conte of the protein usually increases during the early stages of infection. When the effect of the pathogen is seen, little change, or in some cases, a decrease in plant protein has been found in most diseased tissues. In the case of crown gall tumours, there is much more protein in them than could be expected from synthetic activities of the pathogen. In late degenerative stages of disease, big decreases in protein, sometimes associated by enhancement in free amino acids, are often found.

It is impossible to generalise about the influence of virus infection on host-protein metabolism. A few viruses such as TMV and potato virus X, in connection with which this problem has been most intensively studied, may be a typical of plant viruses as a whole; these viruses multiply to such an extent that the virus protein contributes significantly to the total protein of the leaf. The synthesis of virus protein in turn must exert a considerable direct stress on host protein metabolism. Most viruses, however, contribute very little to the total plant protein.

In several virus infections, there may not be very many

total new proteins produced after infection, although in some infections, considerable virus is produced. There is a redirection of protein synthesis. New enzymes are synthesized, and in the case of TMV in tobacco, the ribosomes and plastids of the host degenerate. The host RNA tends to increase after infection, followed by the appearance of viral RNA. In advanced infection of TMV there may be as much as 30% of the total soluble protein of tobacco leaf. Wildman *et al.* (1949) analyzed extracts of infected leaves in the boundary electrophoresis apparatus. They determined that in a given leaf, as the infection progresses, total nitrogen is unaffected but the virus is syntherized at the cost of normal nucleoprotein found in the cytoplasm of the leaf cells. It is generally agreed that there are big differences between the nitrogenous constituents of normal resting cells and actively dividing tumour cells incited to division by the crown gall bacterium *Agrobacterium tumefaciens* Neish and Hibbert have reported that while both normal and tumour tissues contain about the same amounts of non-protein nitrogen, tumours have three times the protein of normal tissues and the difference in water soluble protein is 6:1 in favour of tumour tissues. This was interpreted as a greater tendency and capacity on the part of tumour tissues to synthesize protein.

According to studies by Klein, the biggest increases in proteins and soluble nitro-gen became obvious in tumour tissues late in their developmentent at the time of cell volume enhancement. He showed that increased protein synthesis registered in the tumour tissues at this time did not occur entirely at the expense of free amino acids and amides in close proximity to the tumour. On the basis of these observations of the stunted and nitrogen deficient appearance of tumorous plants, Klein stated that the nitrogen of the whole plant was being mobilized into the tumour.

Walker studied the amino acids and amide pool of the bean plants, susceptible and resistant to the haloblight bacterium *Pseudomonas phaseolicola.* The amino acids and amide content were not appreciably changed in resistant varieties as a result of infection: Chrom atographic analysis of a healthy and inoculated leaf of a susceptible variety revealed 22 ninhydrin positive compounds, and of these, 15 were identified as amino acids and amides.

Changes in amino compounds were determined in inoculated and chlorotic tissues and again there was a large pile up of ornithine as well as histidine and methionine in the rinoculated leaf tissue. More than five times as much ornithine piled up in the Uninoculated chlorotic leaves.

Nitrogen Metabolism of Fungus Infected Plants

Generally in fungus infected plants the total nitrogen and protein capacity of the host pathogen complex increases. In infected beans there is a greater incorporation of cysteine labelled with S^{35} in the protein fraction of the leaf at the infection, sites than in the uninfected parts of the leaf. This shows that there is greater protein production in diseased tissues. The acid soluble fraction of the leaf also increases, which is an indication of the enhanced uptake of cysteine into the leaves. This increase of proteins is mainly due to the production of proteins by the pathogen rather than the host, as experiments done with tritium-labelled glycine showed that it was mainly incorporated into the fungal mycelium and uredospores. Farkas and Kiraly studied that the free ammonia content of leaves increased only in wheat leaves infected with a race of stem rust which resulted in a susceptible reaction. No increased ammonia content was seen in resistant reactions, while partially resistant combinations yielded intermediate amounts of ammonia.

On the other hand enzymes engaged in amino acid and amide synthesis are also activated in rust infected plants; showing an increase of various amino acids and amides. Thus, a parallel proteolysis and protein synthesis can be seen in rust diseases Rudolf compared the nitrogen metabolism of wheat infected with rust *(Puccinia graminis trilici)* in intact (attached) and detached leaves. With infected leaves attached to the plant, a partial proteolysis was experienced, resulting in an accumulation of free amino acids and amides, while a relative increase in protein synthesis was suggested after infection in detached leaves accompanied by a decrease in free amino acids and amides. Detached healthy leaves are normally characterized by a high rate of proteolysis with an accumulation of amino acids and amides, but in this case, infection by the rust pathogen appears to have diminished the observation of this phenomenon. Kiraly reported that in attached leaves of bean infected with rust *Uromyces phaseoli*, there is an enhanced protein production in the tissues of the green islands surrounding the infection site, while the tissue zones between the green islands become aged and are characterized by a low capacity for protein synthesis.

Many investigators have tried to find a relation between the resistance of host tissues and their nitrogen metaoolism. Thus, Lakshminarayanyanan insisted that the ratio of non-protein to protein nitrogen remains unchanged under the effect of infection in cotton varieties resistant to the *Verticillium* wilt, while a big decrease in this ratio was noted in the susceptible varieties. The results in the field of rust diseases indicate that a resistant variety exhibiting an incompatible reaction is characterized by a different nitrogen metabolism than that of a susceptible one. The resistant varieties shows upon infection a decrease in total soluble and protein nitrogen which is in contrast to susceptible varieties.

Enhancement in the size and further breakdown of the nuclei in the infected resistant host cells also occurs. but these are more rapid than in the compatible host plant. Total nitrogen on a dry weight basis is higher in diseased than in healthy roots and although the total of free and combined amino acids is only slightly higher in diseased roots, there is a two fold increase in free alanine, serine and asparagine.

Natsvlshvili observed that lysine, arginine, and methionine were in higher level in maize leaves susceptible to *Helminthosporium turcicum*. Padmabhai reported that rice plants, highly susceptible to helminthosporiose, contained more glutamic acid In rice plants infected with *Gibberella fuzikuroi*, the amino acid proline, disappears as a result of infection Chowdhury and Mukhopadhyay reported that changes in the ribonucleic acid contents in rice leaves was caged due to tungro virus infections. Vidhyasekaran and Parambaramani worked on the nitrogen metabolism of alga infected plants. Algal infections were found to be the cause of reductions in the total protein, ammoniacal, nitrite, amino and amide nitrogen content of mango, guava and *Sapota* leaves, while nitrates amassed in the infected leaves. Glutamic acid and alanine increased markedly, glycine decreased sharply, whereas valine contents were not affected by algal infection due to *Cephaleuros* infection.

Naik and Addy stated that there was no relation between a change in free amino acid contents and infection by *Cercospora personata*. There were no qualitative or quantitative differences between resistant and susceptible cultivars. The amino acid content of bajra in relation to ergot disease resistance was studied by Kannaiyan. They reported the presence of specific amino acids in large quantities in the susceptible varieties. Asparagine and proline which were there in large quantities in the susceptible varieties enhanced the growth of the pathogen *in vitro*. Amino acids which were present in large quantities in resistant varieties have been reported to be inhibitory to the pathogen. Tryptophan, the amino acid present in excess in the resistant spikelet, was inhibitory to *Claviceps microcephala in vitro*. Sekhawat and Kothuri investigated the amino acid concentration of healthy and downy mildew plants of opium poppy and they reported that the concentration of a number of amino acids was either decrease in the samples of diseased plants or it remained save except for tyrosine which was absent and their known amino acids which increase in samples of diseased plants. Vidhyasekaran reported an increase in the protein content of rice after infection by *Helminthosporium*. Citrus fruits, infected by *X. citri*, also showed increases in their protein content.

Arya and Tiwari studied the altered nitrogen metabolism of healthy as well as diseased bajra tissues infected with *Sclerospora graminicola in vivo* and *in vitro* conditions. Free arginine and aspartic acid appeared specific in the infection process. Both the amino acids shows big increase in all the tissues associated with the disease. Serine, glutamic acid and aspartic acid in the protein form were important and associated with the diseased tissue in a decisive manner.

Many efforts have been made to learn something about the nitrogen metabolism of rusted plants by combining inoculations, with substances that changed the course of disease development. Resistance to *Puccinia graminis tritici* is lost when detached leaves are floated on water but not when solutions of 40 ppm benzimidazole are used. The behaviour of the leaves of susceptible plants is not changed. Another striking effect of benzimidazole is that it prevents the concentration of protein that occurs in detached leaves from falling. It may, at the same time, cause a greater resynthesis of proteins and a drop in the levels of soluble nitrogen and alcohol soluble carbohydrates as compared with the leaves floating on water. Other treatments, such as applying maleic hydrazide to the soil before plants are inoculated also called enhancement in the levels of carbohydrate and soluble nitrogen in leaves. Resistance is associated, therefore, with high levels of proteins and low levels of soluble nitrogen and carbohydrates, and susceptibility with the reverse of these conditions.

Vidhyasekaran showed that in *Eleusine coracana* (finger millet), leaves susceptible to *Helminthosporium nodulosum* have more of asparagine, methionine and cysteine than the resistant leaves. Applying of nitrogenous fertilizers enhanced the incidence and contents of these three amino-acids/amides. Asparagine produced more growth of the pathogen *in vitro* as well as the toxin and enzymes producing ability of the pathogen, while methionine and cysteine inhibited it. Methionine and cysteine reversed the inhibitory action of quinones on the activity of pectic and cellulolytic enzymes produced by the pathogen. Thus, it is quite possible that asparagine, by its ability to produce more growth and enable greater toxin and enzyme synthesis, by the pathogen; and mothionine and cysteine, by their ability to reverse the inhibitory action of quinones on enzymes produced by the pathogen, would make the host susceptible to the pathogen.

PHENOLS

Phenols are compounds which have one or more aromatic rings with one or more phenolic hydroxyl groups. The simplest compound is phenol. Compounds having more than one hydroxyl group on the benzene ring are called polyphenols. Phenolic compounds are widely present in higher plants and fungi. Anthocyanins, leucoanthocyanins and anthoxanthins, sugar esters of Quinones and shikimic acids, esters or hydroxycinnamic acids and compose class of compounds are known as phenolics. Recently, the term polyphenol, have been used to refer to these compounds. Phenolic compounds function as hydrogen donors or acceptors in oxidatiop reduction reactions and they play an important role in signification Some phenolics show antiauxin activities and are strong growth inhibitors. Others act synergistically with auxins in growth stimulation. Phenolic compounds have been shown to play key roles in the resistance of plants to disease causing micro-organisms. In addition, phenolics may also act as protectants against leaf eating insects, for example, they may influence the availability or digestibility of proteins which insects obtain from leaves. Phenolics may also intervene with growth and other energy dependent activities by uncoupling oxidative phosphorylation. The oxidation of phenols yields highly reactive Quinones which inhibit enzymes by complexing with metal ions, reacting with sulfahydryl groups or binding nonspecifically to proteins. It is clear that even small changes in phenol metabolism may seriously disrupt many processes which are most for normal plant growth and development.

Evidences from ^{14}C tracer elements indicate that the biosynthesis of phenolic compounds in plants may proceed via one or more of three different pathways.

a. Shikimic acid pathway.

b. Acetate malonate pathway.

c. Acetate-mevs!onste pathway.

The most important of these is the shikimic acid pathway (Figure) in which the initial reaction involves the condensation

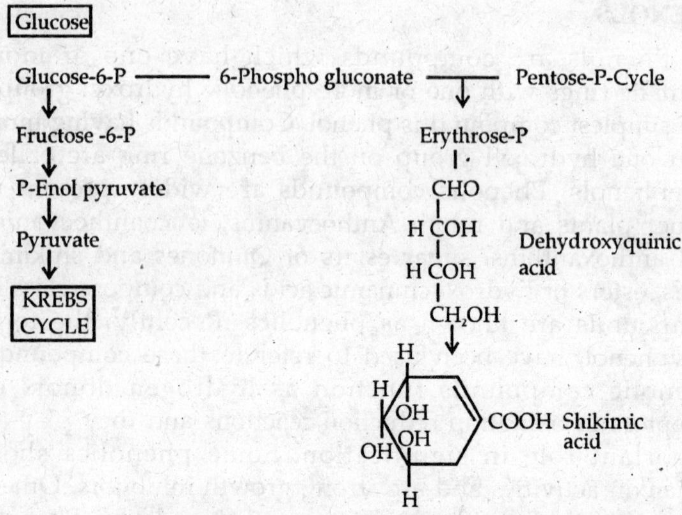

Figure. The shikimic acid pathway.

of phosphoenol pyruvate and erythrose-4 phosphate to yield a seven carbon intermediate which is converted by a series of reactions into shikimic acid.

The Shikimic Acid Pathway

The shikimic acid pathway studied first in bacteria, appears to function in higher plants for synthesis of phenolic compounds. The initial reaction of this. pathway involves the condensation of phospho enolpyruvate and erythrose-4-phosphate.

Phosphoenol pyruvate is supplied by glycolysis, whereas erythrose-4-phosphate is derived from the pentose-phosphate pathway. One reaction, the reduction of 5-dehydroshikimic to shikimic requires NADPH which is also a product of the pentose-phosphate pathway. Transaminations convert shikimic acid to phenylalanine or tyrosine. In dicotyledons, phenylalanine ammonia lyase (PAL) converts phenylalanine to trans-cinnamic acid and thus provides phenyl propane skeletons which can serve as building blocks for lignin or be utilized in the synthesis of flavanoids and various other phenolic derivatives. In cereal grasses, tyrosine ammonia lyase change tyrosine to p-coumaric acid (4-hydroxycinnamic) and this, in addition to the PAL system, gives phenylpropane skeletons which serve the same functions as those provided by PAL. In several plant tissues, ammonia lyase activity has been found to be closely related to the biosynthesis of phenolic compound PAL activity has been shown to increase in response to a number of stimuli, induding light and mechanical injury as well as infections by viruses, bacteria and fungi. The- level of PAL activity increases to a peak within 24 hours of infection by *Ceratocystis fimbriata* and this increase is followed by a rise in the concentration of polyphenols (Figure).

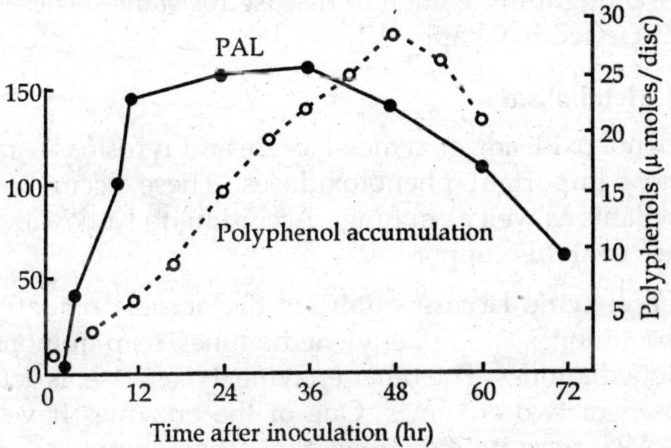

Figure. Activiy of PAL and polyphenol accumulation in discs of sweet potato tissue following inoculation with *Ceratocystis fimbriata.*

Acetate-Malonate Pathway

Some phenolic compounds are produced by head to tail condensation of acetate units. This pathway, known as the acetate malonate pathway, is thought to involve acetyl Co-A and malonyl Co-A as intermediates and resembles fatty acid synthesis. Isocoumarin (3-methyl,. 6 methoxy, 8-hydroxy, 3-4 dihydro isocoumarin) whkh accumulates in carrot slices is apparently synthesized via the acetate malonate pathway.

Acetate-Mevalonate Pathway

In a third pathway, the acetate-mevalonate pathway,

acetyl Co-A units are condensed to form mevalonic acid, which is then changed into cyclic terpenes or steroids. The formation of compounds with two or more aromatic rings often involves interactions among different pathways. For example, both the acetate malonate and shikimic acid pathways contribute to the synthesis of the isoflavone and pisatin.

The current interest in phenol metabolism stems from the identification of a large number of compounds synthesized via the shikimic acid or acetate pathways which accumulate after infection or injury. Many of these compounds, called phytoalexins, have antibiotic properties and are thought to function in disease resistance. These have been discussed in Chapter VI.

Phenol Metabolism

Phenol oxidising enzymes-laccase and tyrosinase, are the two most important phenoloxidases. These occur both in higher plants as well as in fungi. According to Kubowitz both enzymes contains copper.

The enzyme laccase catalyzes the aerobic oxidation of p-hydroquinones or p-phenylenediamines to p-quinones or p-quinonediamines. The other enzyme, tyrosinase, is actually composed of two enzymes. One of the enzymes have only catecholase activity, whereas the other shows cresolase activity. This enzyme (tyrosinase) complex is also known as phenolase, phenoloxidase, polyphenoloxidase or DoP A-oxidase. Tyrosinase is also an aerobic oxidase which is able to oxidize many phenol compounds by oxygen. However, neither para nor meta-dihydroxyphenols are suitable substrates for this enzyme since it is able to oxidize only ortho-dihydroxyphenols.

The enzymatic activity which makes O-diphenol from monophenol oxidatively (monophenol +½ O_2 = O-diphenol) is frequently referred to as cresolase activity. On the other hand, catecholase activity oxidizes the produced diphenol to quinone (O-diphenol + H_2O = O-quinone + H_2O). These activities are given in Figure.

MONOPHENOL + $\frac{1}{2}O_2$ ⟶ ORTHODIPHENOL

ORTHODIPHENOL + $\frac{1}{2}O_2$ ⟶ ORTHOQUINONE + H_2O

Figure. Catechol activity of the phenolase enzyme

O-Quinone produces "melanine", which is a dark-coloured pigment produced in the course of further enzymatic or non-enzymatic reactions. Melanine is considered to be an end product but the details of the formation of this complex are not yet known.

Changes in the activity of phenol oxidizing enzymes, peroxidase and phenolase may also play a role in the regulation of metabolic pathways in diseased or injured tissues. Phenolase is a copper-containing enzyme which catalyses the oxidation of mono- and dihydroxyphenols to phenols. The resistance to several fungal pathogens has been ascribed to a higher concentration of fungitoxic phenolic substances and their oxidation products, to increased phenolic substances and their oxidation products and to increased polyphenol. I oxidase activity which generally, but not invariably results from infection. More than 40 years ago Dufrenoy studied that when the progress of some pathogens is checked, phenolic compounds, mostly tannins of the gallic group, groups in abundance within the vacuoles of adjacent cells, whereas in specialized parasites, tending to become systemic, tannin is produced very slowly in small amounts and only in vacuoles close to the parasite. Subsequently, it was suggested that the establishment of *Puccinia coronata* in oat leaves depended on the supply of phosphorus from the host tissue; on the decrease of this element, relatively innocuous host phenolic compounds were dehydrogated by tyrosinase into toxic quinones, leading to the localized necrosis of the tissue.

IMPORTANCE OF PHENOL OXIDATION IN PLANT PATHOLOGY

The phenol oxidase system plays an important role in respiration by transferring electrons from respiratory substrates to other hydrogen or electron acceptors. Oxidation products of phenols (qui nones) may be reduced to their original phenol form by respiratory carriers. $NaDH_2$ and $NADPH_2$, generated by the oxidation of respiratory substrates, may reduce quinones in the presence of quinone reductase enzyme. Polyphenolases oxidize monophenols to polyphenols through cresolase activity in which the hydroxylation of monophenol leads to the ormation of complicated polyphenols, such as flavonoids, tannins, lignins and others, whereas the catechol activity has a role in the browning reaction of plant tissues.

Phenolase may influence the amino acid metabolism of plants. In many cases a close correlation exists between the increased activity of phenolase and peroxidase and the concentration of phenolic substances. Phenolic substances in turn have been correlated with resistance to pathways.

The enhanced activity of phonolase diseased or wounded areas of plant tissues is generally associated with an increased concentration of phenolic substances. Polyphenols, like chlorogenic and caffeic acids play an important role in the inhibition of the activity of indole acetic acid oxidase. Phenolics and their oxidation products seem to play a role in the auxin metabolism of plants. Polyphenols, such as chlorogenic and caffeic acid, known to be powerful inhibitors of IAA-oxidase, may in turn be able to inverse the enzyme activity and confer resistance via an auxin-spring action on plants that produce them in response to infection.

Oxidized polyphenols inhibit the pectinolytic enzyme produced by various fungi and they have clearly been associated with the resistance of apples to brown rot.

It has been studied in many experiments that a relation may present between the intensity of resistance and the phenol level in healthy plants. Chandramohan showed that varieties of *Amaranthus tricolor* resistant to *Alternaria* sp.

accumulate O-D phenols and the amount of the latter was many times that of the susceptible varieties. Walker and Stahmann showed that onion varieties resistant to *Colletotrichum circinans* amas-flavones, anthocyanine and simple phenolics, such as protocatechuic acid and catechol in the dried bulb scales which are red pigmented.

It has been shown that oxidized phenols as well as oxidized phenolic glucosides are even more better. In the case of apple leaves., resistance to *Venturia inaequalis* has been attributed to Phloretin and that of pea shoots to *Erwinia amylovora* is attributed to arbutin. This is due to the action of β -glucosidases, polyphenol oxidyses and peroxidases.

Lee and Le Tourneau showed the accumulation of higher amounts of chlorogenic acid in the potato roots of varieties resistant to Verticillium wilt disease. Mahadevan showed that phenols and their oxidation products showed inhibitory activity on the pecto lytic and cellulolytic enzymes of the pathogen, *Cladosporium cucumerinum.* He suggested that the defense mechanism in resistant plants should involve the importance of phenolic constituents and their oxidation products on the cell wall degrading enzymes of the parasite. The inactivation of extracellular microbial enzymes by the phenols or Quinones present in plants after penetration is certainly a defense mechanism.

According to Johnson and Schall, the periderm of tubers from a variety resistant to scab showed much more chlorogenic and caffeic acid and were largely limited to the outermost tissues where the scab organisms proliferate. In the Verticillium wilt of potato, the resistance is attributed to a greater quantity of chlorogenic acid in root tips, sprout tips and the vascular system of resistant varieties.

In the case of *Phytophthora infestans* infection, the leaves of resistant potato varieties show a more intense accumulation of toxic oxidative derivatives of polyphenols than do the leaves of susceptible varieties. According to Tomiyama, the resistance *to Phytophthora* is reversed by copper enzyme inhibitors that interfere with the formation of toxic quinones via polyphenol oxidases (PPO).

Following inoculation with *Phytophthora infestans,* both polyphenol and PPO increased in resistant potato varieties but not in susceptible varieties which proves a resistance mechanism to the enzyme. Voros have reported that when streptomycin is absorbed through the root, it gives resistance to potato *againstPhytophthora.* This is due to the fact that antibiotic activity generally enhances PPO activity, and hence, the resistance in plants.

The activity of PPO would seem to be important in that it can oxidize phenolics to quinones which may be more fungitoxic. Polymerisation of quinones to dark coloured tannins may occur. In tomato wilt conjugated phenols such as glucasides, tannins and lignins may be hydrolyzed by enzymes to toxic qui nones which then polymerize to cause characteristic vascular discolouration as suggested by Davis and Dimond. PPO produced by the pathogen might oxidize host polyphenols to more highly fungitoxic substances which prevent further development of the fungal pathogen. For example, a PPO inhibitor reduces the toxicity of catechol to *Cochliobolus miyabeanus* and prevents the development of melanin presumably by preventing enzymic conversion of catechol into fungitoxic quinones. The melanin is fungitoxic although in other cases these insoluble pigments are relatively non-toxic to fungi. Such inhibition can thus result from the production of PPO by the pathogens.

The germinating uredospores of *Puccinia gramins produces* phenols and PPO, and the resistance of barley plants to *Erysiphe graminis* appears to be associated partywith an accompanying production of a phenolic substance which accumulates around the haustoria and inhibits further development of the fungus. Hare points out that aromatic substances, such as polyphenols, phenolic glucosides, flavonoids, anthocyanins, tend to accumulate in and around infected plant tissue and also in tissue near to wounds, where presumably they might exert a fungitoxic effect. Kuc' described two types of reactions in apple varieties resistant to *Venturia inaequalis.* Phloridzin, a polyphenolic glucoside present in tbe bark and leaves of apple, was hydrolyzed to glucose and phloretin by a β-glycosidase and the phloretin was oxidized by PPO to a fungitoxic intermediate compound and later to melanins. These intermediate compounds, possibly quinones, might responsible for resistance to the fungus. They are release in susceptible and resistant apple leaves but their production is much quicker in the latter, thus leading to rapid inhibitjon of the invading pathogen.

Kuc' suggested that a great variety of phenolic compounds present in plants might provide a partial explanation for the specificity of the host-parasite interaction. For example, certain plant species (resistant) might contain or form phenolic substances toxic to a special pathogen, whereas others (susceptible) might lack them.

GROWTH REGULATORS IN PLANT PATHOLOGY

Symptoms, such as galls, overgrowth or malformation of host tissues, epinasty, elongation, witches broom effects, etc. shows abnormal growth activities in many plant diseases. The extracts of tissues which shows such abnormal growth patterns are often ten times more active in the tests, for growth-regulating substances than are the extracts of healthy plants. Even in the absence of obvious growth abnormalities; diseased tissues often have increased levels of growth regulators. Four classes of compounds are known to be involved in plant diseases. All the major types of plants growth regulators (auxins, gibberellins, cytokinins and volatile hormone ethylene) are produced by various viral, bacterial and fungal pathogens. An study of the role of individual growth regulator in pathogenesis is complex because their physiological effects in many cases overlap. Growth regulators in plant diseases have been reviewed by Sequeira.

Auxins

The auxin naturally present in plants is Indole-3-acetic acid (IAA). It is continually produced in growing plant tissues, in which, it is present in concentrations of about 5

$\mu.g$ per kg of green plant material. β -3gudole Acetic acid) perhaps the nearest to being ubiquitous. Auxin is an organic substance, that promotes growth along the longitudinal axis when applied in low concentrations to the shoots of plants which are freed, as far as possible, from their own inherent growth promoting substances.

"Went", first studied the effects of IAA on *Avena*, following which it was shown that IAA has a widespread presence in higher plants and especially in diseased tissues. Auxin is also synthesised freely by many species of bacteria and fungi. The synthesis of IAA in very small amounts takes place in young leaves and root tips. IAA is required for cell elongation and differentiation. Cell elongation is thought to be brought about by the removal of Ca++ and Mg^{++}cation bridges from the pectate network which leads to the formation of straight chain pectic acids or pectinic acids and this in turn allows greater plasticity and extensibility to the cell wall. IAA also facilitates this by forming a complex with, and thus inactivating the enzyme pectin methyl esterase (PME) on the cell membrane and so further stoping the formation of sites suitable for the attachment of divalent cations. The absorption of IAA to the

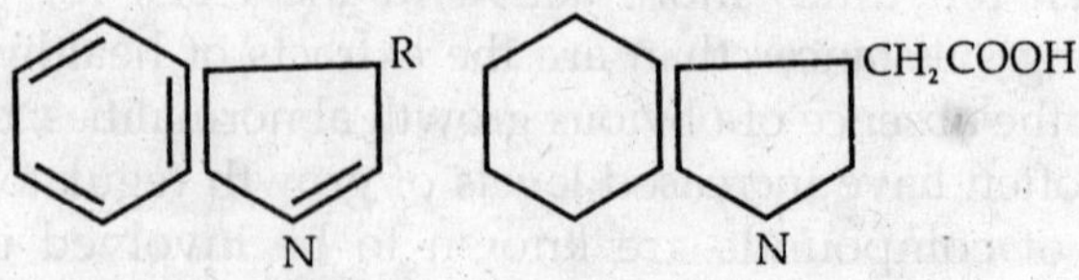

The indole nucleus Indole-3-yl acetic acid (IAA)

cell membrane also affects the permeability of the membrane. Many basic functions have been attributed to IAA. It has been linked - with cell metabolism, phenol metabolism, increase in respiration, enhancement of the pentose phosphate pathway activity and membrane permeability. Evidence has also accumulated of the primary site of action of IAA being on the nuclear control of RNA synthesis or on RNA itself, thus affecting the level of certain key enzymes in cell growth.

Viral Disease

"Grieve" attributed the reduction in the growth of

tomato plants infected with tomato spotted wilt virus to a decrease in auxin content and to decreased rate of translocation through the stem. Surprisingly, Jones did not find a decreased auxin content in the tomato tissues infected with spotted wilt virus. This can be explained by the observation made by Best that such infected plants contain a high level of scopoletin which is a known inhibitor of IAA oxidase. Symptoms of dwarfing in many viral infections are due to the lower-auxin content of the tissues.

Bacterial Diseases

The synthesis and role of auxins in plant diseases has been studied in detail in some bacterial diseases of plants, such as *Agrobacterium tumefaciens, Pseudomonas savastoni* and *Pseudomonas solanacearum*. One of the best known plant diseases, displaying overgrowth symptoms, is crown gall caused by *Agrobacterium tumefacie.ns*. There is enough evidence that the large amounts of auxin found in crown gall tumour are produced by the host. This disease has been studied by many scientists as a model system in cancer and tumour genesis. Among other things, the pathogen evokes symptoms by altering the hormonal balance of the host cells. Although auxin is produced by *A. tumefaciens*, the non-auxin producing strains of the organism are also pathogenic and in nature cause galls to be formed. Thus, the pathogen is able to invade the host to synthasize abnormally high concentrations of auxin. This is accomplished by a pathogen produced factor known as the tumour inducing principle (TIP). While the normal cell is being changed into a tumour cell, its growth increase in an uncontrolled manner and becomes independent of the bacteria. The identity of the transforming agent, TIP, is unknown. TIP has been suggested to be the bacterium, a bacteriophage (a virus associated with the bacterium), a metabolic product of the bacterium, a converted host component, or more recently, bacterial DNA. Although the recent hypothesis that transformation may result from the uptake, incorporation and transformation of a part of the bacterial genome by host cells is attractive, but there is no proof. Equally uncertain is the role of undefined cell division which promotes the factors reported to be present in crown gall tissues.

In main, there is little evidence that the auxins produced by the pathogens play a main part in studying pathogenesis or in determining the nature of plant pathogen interaction. Instead, changes in auxin activity during pathogenesis probably result from alterations in host metabolism, produced by the pathogen.

Fungal Diseases

There are many reports about the production of IAA in bulk from culture media supplemented with tryptophan or other IAA precursors. This has been used as more than circumstantial proof for the production of IAA *in vivo* by the particular pathogen.

"Shaw" and "Hawkins", found substantial increases (24-times) of auxin corresponding to IAA in Little Club wheat susceptible to and infected with *Puccinia graminis tritici.* In the case of Atlas barley susceptible to *Erysiphe graminis,* on the other hand, infected plants showed a maximum five-times increase in IAA in the infected, leaves as compared with the controls. Rust-infected cereals, unlike other rust-diseased hosts, do not show hyperauxiny. enhanced auxin activity has bem demonstrated in the infected tissues in the aecidial stages of *Gymnosporangium* on *Pyrus* and *Uromyees* on *Ranunculus.*

The auxin content and auxin catabolism of the stems of *Eupltorbia cyparissias* infected with *Uromyces pisi* were studied by Pilet. He showed that IAA concentrations in the host are universali associated with the IAA oxidase activity of the host. The inhibition of IAA oxidase may result in the accumulation of auxins in the lesions.

Auxins are also involved in vascular wilt diseases. Beckman showed that cultures of the fungus synthesizes a substance similar to IAA. It was claimed that the synthessizes of this substance *in vivo* increased the plasticity of xylem parenchyma walls allowing a pectic enzyme weakening of pit membranes and resulting finally in tylosis. IAA solution induced the formation of tylosis in tomato shoots. This supported the view that, auxin is involved in the disease syndrome of the wilt. Increased levels of IAA have been reported in infected stems. Auxins have also been implicated in resistance to ascular wilting. According to Waggoner and Dimond and Corden and Dimond, ethylene, produced in the infected tissues, gives protection from infection.

Vidhyasekaran and Durairaj, discussed the possible reasons for the shot hole symptoms caused in mango leaves, by the pathogen *Colletotrichum gloeosporioides*. Indole acetic acid and phenols amassed in the lesions, while there was a reduction in IAA content in the tissues around the lesions. Both IAA oxidase and peroxidase activity was inhibited in the lesions while there was an increase in polyphenol oxidase. According to them, tissue breaking is due to the accumulation of auxins in the lesions. The zones around the lesions have less auxin, and hence, uronic carboxyls predominate and the cell walls are rigid. The torn tissue in the lesion *per se* slough out, leaving the surrounding tissues rigid.

Gibberellins

Gibberellins are the group of chemically similar growth-regulating substances. They can be considered to be normal components of green plants. The existence of gibberellin in plants might still have been unknown if the "bakanae" or "foolish seedling" disease (which had devastating effect on the rice economy of Japan) had not broken out. Japanese farmers noted that plants affected with this disease out-grew normal plants but were thinner and paler than their normal counterparts and there were sometimes lack of fruits. In the early part of this century, Japanese pathologists first showed the connection between the bakanae disease and the fungus called *Fusarium moniliforme* Sheld.

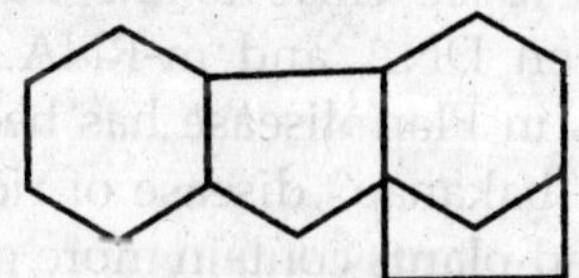

Gibbane skeleton

Sawada proposed that disease may be caused by some

product of the fungus. Yabut and Hayshi (1932) were able to seperate the crystalline gibberellin from the infected rice plants. Since that time, gibberellins and alike substances have been found in higher plants and implicated in disease symptoms. All these compounds possess the same general carbon skeleton and all are able to promote either stem elongation or cell division or both in plants. They have a gibbane skeleton.

Numerous functions have been attributed to gibberellins. Thus, gibberellin is the chemical signal, that is, the hormone secreted by the embryo in the seed that catalyses the cells of the aleurone layer into secreting hydrolytic enzymes, such as amylase, which is responsible for liquifying the reserve starch. It also activates the enzymes that promote cell wall break down, thereby aiding in the digestion of the endosperm cells and in weakening the seed-coats which further allows the seeds to germinate. Besides inducing amylase activity in the aleurone layer of seeds, gibberellins exert a potent stimulus on terminal meristems, and thus in some respects, mimic the effects of auxins on plant growth. Gibberellin also activates or induces the synthesis of other enzymes, including proteinases. Proteinases break down the reserve proteins of the endosperm and release, various amino acids, including tryptophan. Tryptophan is then activated to form auxin (IAA).

Gibberellins, generally have striking growth-promoting effects; they speed the elongation of dwarf varieties to normal size, promote flowering, elongate the stem and root and encourage the growth of fruit. Auxins and gibbe: rellins may also act synergistically. The point of action of the gibberellins in the cells'seems to be close to the nucleic acid system, somewhere between DNA and m-RNA. The involvement role of gibberellins in Plan disease has been studied in few diseases. In I. the "bakanae" disease of rice, it is not known whether the infected plants contain more gibberellin than the healthy plants, or in what manner the fungal gibberellin causes the sym. ptoms observed. Following the classical discovery of the "bakanae" disease, the first truely experimental investigations on endogenous gibberellins and infected plants were made by Radley. It was shown by her that yea nodules infected with *Rhizobium* contain an increased level of GLS as compared to healthy roots.

Gibberellins have been studied more extensively in relation to viral diseases, although no difference has been reported so far in the gibberellin contents of healthy and virus-infected plants. The spraying of diseased plants with gibberellins has been shown to overcome some of the diseases caused by viruses. Thus, the stunting of corn plants infected with corn stunt virus (Maramorosch, 1957) and tobacco plants infected with severe etch virus (Chessin, 1957) was reversed after treatment with gibberellin. Axillary bud suppression caused by sour cherry yellows virus (SCYV) on cherry (Hull and Klos, 1958) and leaf curl virus on tobacco; (Nariani, 1963) was also overcome by gibberellin sprays. The fact that stunting, caused by some viruses, can be reversed by applying exogenous gibberellins suggests that the reduced growth of these virus-infected plant might be due to decreased gibberellin content.

Cytokinins

Cytokinins are the strong growth factors which are must for cell growth and differentiation. The first cytokinin discovered was kinetin by Skoog and Miller in the early 1950s. The main action of cytokinin, is on cell division. It is also capable of orienting plant growth induced by auxin. According to Osborne, cytokinins stimulate protein synthesis at the level of DNA dependent RNA synthesis. Cytokinins also delay death in plant tissues. The action of cytokinins is dependent on auxins. They are derivatives of adenine. Kinetin (6 furfurylaminopurine), a compound, has been isolated from yeast DNA by Miller. Kinetin was, however, isolated from herring sperm DNA and has not yet been shown to occur naturally in green plants. Several other adenine derivatives with cytokinin like activities have been synthesized and recently, similar compounds, that is, zeatin and N^6 (2-iso pentenyl) adenosin (IPA) have been isolated from young corn seeds and from spinach and peas respectively.

Cells in cultures show no response when they are supplied with kinetin alone, and they increase in size but do not divide when supplied with only IAA. When both substances arc given together, the cells enlarge and divide. The relative concentrations of kinetin seem to determine the organs that will differentiate, low concentrations inducing root formation, higher concentrations inducitig bud formation. The mechanism of action of cytokinins is not yet known but the available evidence shows that it is similar to that of gibberellins. Cytokinins act at the stage of m-RNA synthesis, possibly by preventing genes from being turned off and by activating genes that have been previously turned off. The role of cytokinins in plant disease has only now become a subject of study. A six time increase in cytokinin activity was detected in extracts from the galled roots of root knot nematode-infected tobacco plants compared with extracts from healthy plants; the cytokinin, however, appear late in the infection since they can be detected in galled roots 75 days, and not 35 days, after inoculation. Root cells near a nematode's head display hypertrophy with repeated nuclear divisions together with the incorporation of neighbouring cells, and this results in large, thick-walled multinucleate syncytia or giant cells. The endogenous levels of several hormones are reported to increase in several hosts infected by nematodes. Cytokinin activity also increases in rust-infected beans and broad bean leaves and seems to be related to both the juvenile features of the green islands around the infection centres and the senescenice outside the green island. In the *Helminthosporium* blight disease of oats, cytokinins do not enhance the toxicity of toxin victorin but they increase the quantity of toxin absorbed. It appears that cytokinins may increase the effectiveness of the pathogens that produce pathotoxins. On the other hand, cucumber leaf discs treated with kinetin become resistant to the powdery mildew fungus, and tobacco leaves infected with the wild-fire toxin and treated with kinetin fail to develop the typical toxin induced chlorosis. A cytokinin has been implicated in the initiation and development of the fasciation of peas caused by

Corynebacterium fasciens. Culture extracts of the pathogens and of infected tissues had materials with cytokinin-like properties. The main active component obtained from the bacterium is, 6-(3 menthyl 2-butanylomino) purine, the riboside of which is found in t-RNA of many bacteria. Plants treated with a high concentration of kinetin attain symptoms similar to those infected by the bacterium. Though these findings suggest that cytokinins play a major role in this disease, a low amount of cytokinin produced by the bacterium in culture indicates that other factors are also involved. Wilson has suggested that cytokinins may be involved in the cell proliferation stage of tumour development in oleander infected with *Pseudomonas savas tanoi*.

Studies on kinetin and plant virus multiplication have been stimulated by the work of Richmond and Lang, and later by Osborne and Moths concerning the effect of kinetin on protein and nucleic acid metabolism and on amino acids accumulation in leaves. It will be recalled that kinetin tends to stop the breakdown of proteins and nucleic acids which occurs when a leaf is removed from a plant and it tends to mobilize the accumulation of amino acids at the point of its application to a leaf. Osborne has suggested that the new RNA and proteins are even produced under these conditions which, without kinetin, greatly favour catabolism.

There are various conflicting reports of the role of cytokinin in the green island formation in infected leaves. Cornu for the first time studied green island formations in the infections of mildews, downy mildews and rusts. Allen pointed out that the green island formation in mildew infected wheat leaf does not merely represent retention of chlorophyll; they actually are areas in which chlorophyll was destroyed and formed subsequently. The green island formation in Pinto bean leaves infected with *Uromyces phaseoli* depends on light conditions after the inoculation of the plants. The total green area which finally becomes visible is the result of a arrest in chlorophyll breakdown and the reformation of chlorophyll. Brian speculated that the green island formation may be due to a release of cytokinin by the pathogen. This is, however, doubtful. Most probably the increased cytokinin level starts from host cells and not from the pathogen. Atkin and Neilands (1972) have speculated about the possible role of siderochromes in the formation of green islands.

An increase in cytokinin content has also been noticed in tumours caused by *Usti/ago zeae* (Turian and Hamilton, 1960) imd *Synchytrium endobioticum* (Reingard and Pashkar, ~958), and peach leaves infected with *Taphrina deformans.* However, peach leaf curl symptoms cannot be entirely attributed to changes in cytokinins, as the level of IAA also increases in infected leaves.

Ethylene

Ethylene (C_2H_4) is a plant growth regulator which is biologically highly active. Ethylene is universally present in plant tissues and is synthesized in increased amounts in damaged or diseased plants. Applied exogenously, it is effective in very low concentrations (below 0.1 ppm) and produces a number of plant responses. Some of these response-epinasty, leaf abscission, premature ageing, fruit ripening and shedding of leaves are similar to the symptoms of certain plant diseases. A number of bacterial or fungal pathogens synthesize ethylene and in the case of fusaria! wilt of tomato caused by *F. 'oxysporum* f. sp. *lycnpersici,* ethylene production by the pathogen is sufficient to account for the epinastic symptoms of a disease. According to Jackson, leaf yellowing and epinasty in waterlogged plants is due to ethylene. Leaf abscission of rose infected with *Diplocarpon rosae* has been described due to ethylene. This is also connected to auxin action in causing tissue proliferation, increased respiration arid stimulation of root formation. The action of ethylene is due to a change in the permeability of lipoprotein membranes.

Muller demonstrated the production of epinasty by emanations from normal and decayed. Non-virulent isolates of C. *fimbriata* could induce resistance and it was claimed that more ethylene was synthesized by them than by the virulent isolates. Sweet potato slices exposed to ethylene (8 ppm for two days) were reported to be rendered resistant to infection and peroxidase and polyphenoloxidase enzyme activity increased during this period. Low concentrations of ethylene were also reported to cause big enhancement in the activity of phenylalanine ammonialyase, a key enzyme in the synthesis of antifungal and antibacterial substances. The enzyme phenylalanine ammonialyase has been implicated in the production of phytoalexins in pea tissue. According to Daly, both the compatible and incompatible combinations evolve higher amount of ethylene from wheat leaves infected with *Puccinia graminis* that control leaves, but the most dramatic increase occurs in the compatible reaction at the time when the pathogen starts to sporulate. In the, case of barley leaves infected with *Erysiphe graminis* a somewhat similar pattern is observed. Peroxidase activity increases in both resistant and susceptible hosts.

Erwinia amylovora does not synthesize ethylene in culture but when cauliflower florets are inoculated with this bacterium, it stimulates ethylene production. This shows that the evolution in diseased tissue most probably is of host origin arising as a result of cellular damage in a still-living tissue. A pectate lyase in the culture filtrate seems to be responsible for the increased production by the plant.

Ethylene undoubtedly interacts with other growth regulating substances, and Van Andel and Fuchs (1972) have pointed out that changes in the balance of growth regulators as a result of such constantly changing interactions may be more important in symptom development than the quantity of any single part. It must be emphasized that it is still not clear whether ethylene is a primary factor in triggering metabolic changes or it is merely produced as a result of these changes.

6

Role of Enzymes and Toxins in phytopathology

ENZYMES IN PLANT PATHOLOGY

For a successful infection by a parasite it should enter the body of the plant from which it will get its nutrition. Hydrolytic enzymes are very important in pathogenesis, since they give the pathogen chemical means or invasion into the host and a process whereby nutrients can be digested. The plant cell is surrounded by a cellulose wall which can be equated to a closed box. If the parasite is to feed successfully, it should break this boundary which acts like a fortress wall. Decomposition and breaking down of host tissues are to be constant components of plant disease syndrome. Enzymes and toxins are the main aspects under which the mechanism of disease has been studied.

The idea that cell wall decomposing enzymes may be involved in pathological manifestations produced in plant tissues by biotic agents was introduced for the first time by De Bary. He explained the action of pathogens on host in two ways one initiates the dissolution of certain constituents of cell walls and subsequently the loss of tissue coherence, and the other kills the protoplast itself. The former effect was traced to some heat labile metabolites possibly enzymes. The second effect could be due to the cell wall degrading enzymes themselves or it may be brought about by a second factor possible toxins. Now a days, pectolytic and cellulolytic enzymes are implicated almost routinely as a feature of host-parasite interactions and their involvement in the degradation of the pectin and cellulolytic constituents of cell walls and of the middle lamella in plant tissues has been reported for such diverse types of diseases as soft rots, dry rots, charcoal rots, damping off, wilts, blights and leaf spots. However, the ability to produce these enzymes does not show that an organism can possess pathogenic abilities. In most plant diseases which are caused by microbial agents, cell walls are penetrated, tissues are colonized and permeability of host cells is changed.

CONSTRUCTION OF CELL WALL MATERIALS AND MIDDLE LAMELLA; CUTICULAR WAX

Plant waxes grows as granular or rod-like projections or as a continuous layer outside or within the cuticle of many aerial plant parts. Formation of wax is a continuous process in the development of a leaf. Recent studies show that minute pores exist below each wax projection and it is by these minute ores that wax is secreted. Cuticular waxes is composed of long

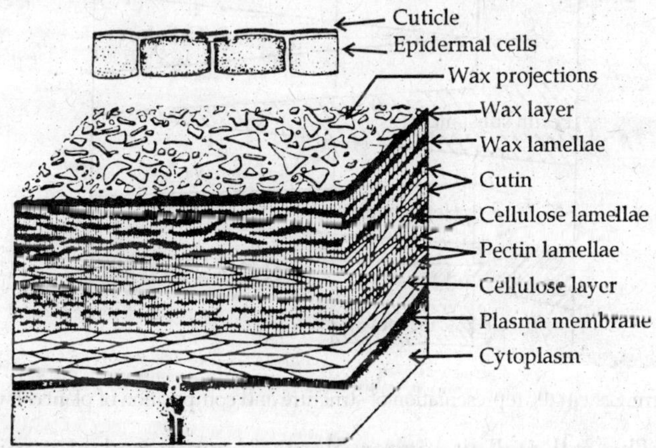

Figure. Distribution in the plant cell wall of the major wall constituents.

chain molecules of paraffin hydrocarbon, alcohols, ketones,

esters and acids. They differ from fats in that glycerol is replaced by monohydric alcohols.

Cutin

Cutin is the main part of the cuticular layer. The upper part of the layer is intermingled with waxes, while in its lower part in the region where it merges with the outer walls of the epidermal cells, cutin is admixed with cellulose and pectin. Cutin, is a polymer of hydroxylated octa and hexadeconic acids. Each chain of the polyester contain 16-18 carbon atoms and two three hydroxyl groups. The hydroxyl groups of one chain are connected to the carboxyl or hydroxyl groups of other chains which gives a three dimensional complex polymeric structure [figure].

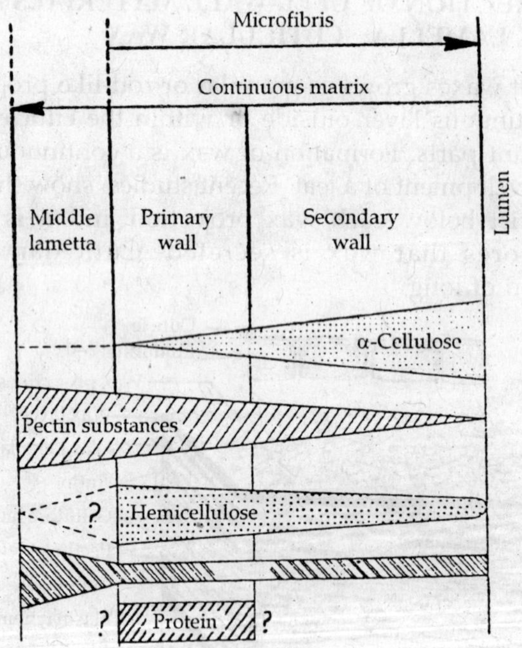

Figure. Schematic representation of structure and composition of plant cell walls.

The cell wall of higher plants is composed of a middle lamella and primary cell wall, and in some cases, a secondary cell wall. Pectic substances are the basic component wall layers and form almost the entire middle lamella, whereas in the other layers, cellulose is also found in good amounts. Beside these two constituents, other constituents are hemicelluloses, lignins, proteins and some other material. According to functions, the cell wall can be divided into three regions middle lamella, primary wall, and secondary wall. The middle lamella acts as an "intercellular cement" that join cells together in the tissue system. A representation of the structure and composition of a typical cell wall of higher plants is given in figure.

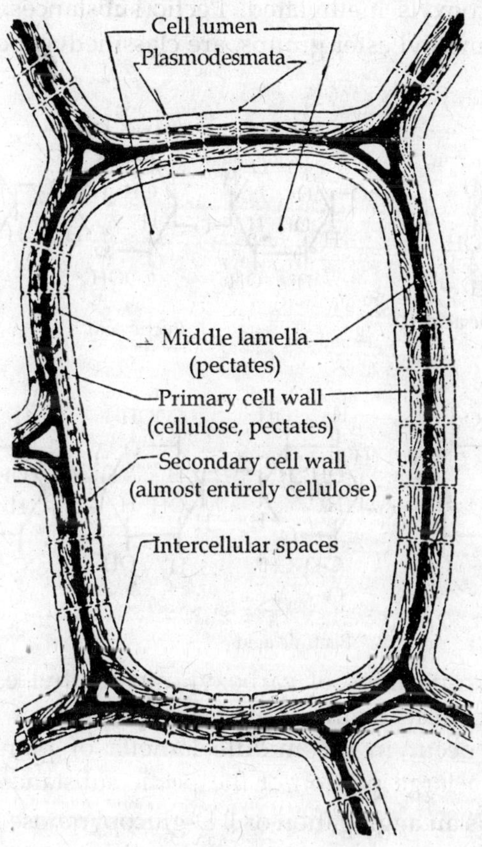

Figure Schematic represention of structure and composition of plant cell walls. (From G.N. Agrios, *Plant Pathology*, Academic Press, London, New York, 1969)

pectius are a group of complex, conoidal carbohydrate derivatives which occurs in or are prepared from plants and the basic structure of which is a linear polymer of Dgalacturonic acid units in a 1-4 glycosidic linkages esterified, with a methyl group or linked with other carboxyl groups ill calcium or magnesium salt bridges. The term *protopectin* is given to substances which does not dissolve in water and which, on restricted hydrolysis, yield pectinie acid. Pectinic acid contains a variable quantity of methoxyl groups, though less than pectin, which may have 75% or more of the uronic carooxyls methylated. Pectic substances, essentially free from methyl ester groups, are classified as pectic acids.

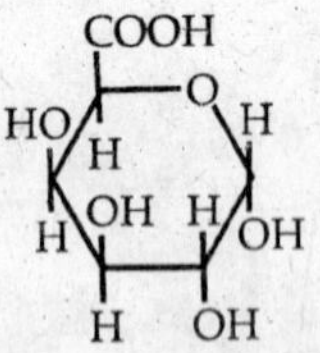

d-galacturonic acid

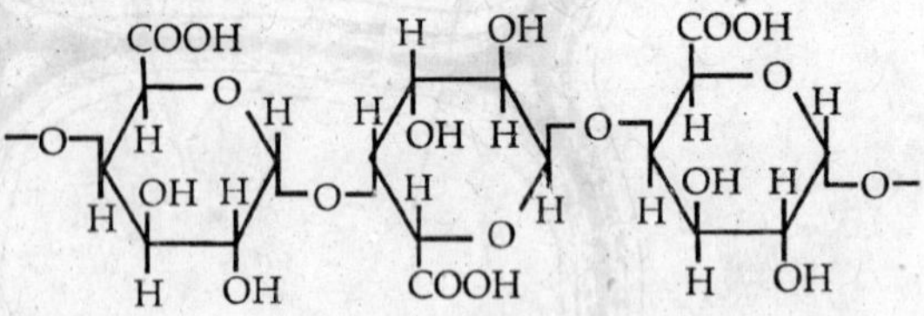

Pectic acid

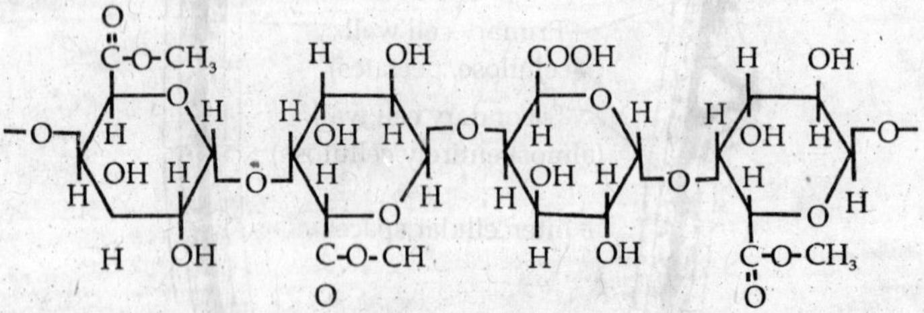

Pectinic acid

The proportions of carboxyl and methyl ester groups have great effects on the properties of polymers. Calcium salts of pectin occur in the middle lamella of plant cells and represent a large portion of the pectic substances in plants. Cellulose is an aggregation of β D-glucopyranose from where C_1 is linked to C_4. Cellulose is the major constituent of the eel walls of higher plants. Cellulose is an insoluble, crystalline substance in its native form [figure].

Cellulose

Structure of cellulose showing β -D-1, 4, linkages of glucopyranose form

Hemicelluloses

Hemicelluloses are also present. With celluloses, hemielluloses are comprised of several polymers. Xyloglucan is an important example of hemicellulose in the primary wall

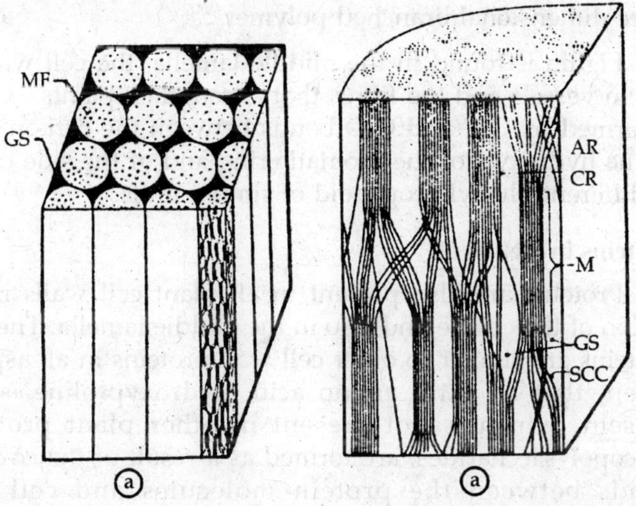

Figure. Schematic diagram of the gross structure of cellulose and micro fibrills (a and of the arrangement of cellulose molecules within a microfibril (b) MF=microfibril, GS=ground substance (pectin, hemicelluloses, or lignin), AR= amorphous region of cellulose, CR=crystalline region of cellulose, M=micelle, SCC=single cellulose chain (molecule).

region. It is made up of a β -1, 4 linked glucopyranose chain with terminal branches of 1,6 linked D-xylopyranose. Other hemicelluloses are glucomannan, a hetetopolymer of D-gluco, pyranose and D-mannopyranose linked β -1, 4 and glactoglucommannan, a glucomannan with 1,6 linked D-gluctopyranose side branches.

Lignin

Lignin is an amorphous polymer having complex structure and high molecular weight. It is different from carbohydrates and proteins. Its structure is still unknown but the basic unit which forms a polymer is phenylpropanoid.

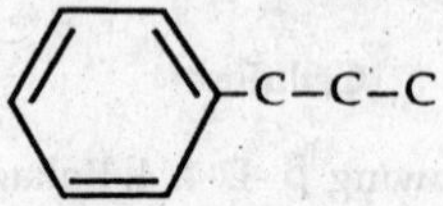

$$C_6H_5\text{–C–C–C}$$

Phenylpropanoid unit

Lignin is produced, from the oxidation products of sinapyl-confiferyl and p-hydroxycin namyl alcohols which condense by free radical mechanism to form an amorphous, three dimensional branched polymer.

Lignin is found in the middle lamella, the cell walls of xylem vessels and the fibres that strengthen plants. Lignin is formed by C-C and C-O bonds between the carbon and/or the hydroxyls of the aromatic ring and of the side chains of different phenyl propanoid of similar units.

Proteins in Cell Wall

Proteins are also present in all plant cell walls in the region of the cuticle and also in the middle lamella. The wall proteins are similar to other cell wall proteins in all aspects, except that an extra amino acid, hydroxyproline is also present, which is not present in other plant proteins. Mucopolysaccharides, are formed as a result of the covalent bonds between the protein molecules and cell wall polysaccharides

Lipids in Cell Membranes

Phospholipids and glycolipids with proteins, are the main component of all plant cell membranes. The common characteristic of all lipids is, that they contain fatty acids that may be saturated or unsaturated.

Enzymes for Waxes and Cutins

Not a single pathogen can produce an enzyme which

can break down wax layers. Wax layers are broken down, waxes cannot be degraded by enzymes.

The significance of cuticle softening enzymes in plant diseases is not certain and needs more studies. The cuticle consists of a cutin matrix with waxes embedded within it and extruded from its surface to give a water proof 'bloom'. Until 1961, no enzymes were known which could degrade cutin. Heinen and Linskens showed that the fungus *Penicillium spinulosum* from rotting leaves produced at least two types of cutinolytic enzymes.

(a) Cutin esterase – This catalyzes the hydrolysis of ester bonds occurring between free hydroxyl and carboxyl groups of cutin bonds.

R-C=O
O
R-C-C-O- (H, O) —Cutin esterase, H_2O→ R-COOH-R-CH(OH)-C-O- (=O)

(b) Carboxycutin-peroxidase – This catalyzes the hydrolysis of the peroxide group of cutin.

H_3C-$(CH_2)_n$-CH-$(CH_2)_n$-COOH
O
O
O=C-$(CH_2)_n$-CH-CH-$(CH_2)_n$-CH_2OH (OH, OH)

—carboxy cutin peroxidase, H_2O→

H_3C-$(CH_2)_n$-CH-$(CH_2)_n$-COOH (OH, OH)
H_3C-$(CH_2)_n$-CH-CH-$(CH_2)_n$-COOH-½O_2 (OH, OH OH)

PECTIC ENZYMES

Pectic enzymes can be classified into two groups (a) Pectin esterases (PME) and (b) Polygalacturonases (PG)

Pectin esterases (PE or PME)

Pectin esterases catalyze the hydrolysis of methyl ester groups of pectinic acids to methyl alcohol and pectinic acids of reduced methoxy content and eventually to pectic acid. PME is found in most higher plants and in micro-organisms.

The microbial PME and plant PME are different in their chemical and physical properties.

COOCH, H OH COOH H OH
-O H OH H O- PME -O H OH H O-2CH,OH
H OH H O H H,O OH H O H H
H OH COOCH, H OH COOH

Polygalacturonases (PG)

Polygalacturonases are chain splitting enzymes which degrade or break down the links between adjacent galactuuronic acid units in pectic substances. This can be done by a hydrolytic mechanism (hydrolases) or an eliminative mechanism (lyases).

The production of cellulolytic enzymes *in vitro* is often reduced by sugars and by various inhibitors, such as polyphenols.

COOCH, H OH COOCH, H OH
-O H H H OH H O- Polygalacturonase H C H H OH H O-
H OH H O H H,O OH H OH OH H H
H OH COOCH, H OH COOCH,

O C-O-CH, H OH O C-O-CH, H OH
-O H H H OH H O-Pectin Lyase -O H C H + H OH H O-
H OH H O H O H PTE H OH H OH H O
H OH C-O-CH, O H OH C-O-CH, O

There is increasing evidence in studies that a number of fungi, particularly the wood rotting fungi, produce several cellulolytic enzymes. *polyporus versicolor* produces a multiple component cellulase system.

HEMICELLULASES

Several micro-organisms – saprophytic and parasitic – create hemicellulases and can change hemicelluloses into pentoses and uronides. Little studies have been done on hemicelluloses as compared with other enzymes now, many scientist are involved in studies of hemicelluloses.

LIGNOLYTIC ENZYMES

Lignin is a resistant material and only about 500 species of basidiomycetous fungi are able of degrading it. Bacteria don't seem important in the break down of lignin. The action of enzymes which decompose lignin have not been properly understood. It has been recognized for many years, that the white rot fungi, together with carbohydrates, decompose lignin in plant tissues showed that *Polyporul versicolor* can destroy over 97% of the lignin in sweetgum wood and many reports of the removal of most of the lignin from wood by white rot fungi is found in the literature. There are several hundred species belonging to a number of families of Hymenomycetes, namely, Agaricaceae, Corticiaceae, Hydnacece, Polyporaceae and Thelephoraceae that can bring about white rot. In addition, a few wood inhabiting Ascomycetes in the order Sphaeriales cause a white rot type of decay

It appears that white rot fungi secrete an extracellular Polyphenol oxidase, possibly lac, case, which enables them to use lignin. On the other hand, all types of wood rotting fun seem to produce *transmethylase* that removes methoxyl groups from lignin, thus "modifying" lignin and increasing the exposure of the embedded cellulose and other polysaccharides to the action of fungal enzymes. Then there are brown-rot fungi, like most white rot fungi, they are members of Holobasidiomycetes and unlike white rot fungi, which decompose both carbohydrates and lignin in wood, brown rot fungi decompose the carbohydrates with only a slight decrease in lignin content.

PROTEOLYTIC ENZYMES

Cell walls have a protein rich in hydroxyproline. Pathogens are known to break down proteins enzymatically, since no particular enzyme responsible for the break down of cell wall protein is known, it is presumed that its degradation follows the same produce as the degradation of cytopasmic proteins. Some of the pathogens which are known to produce the cell wall dissolving enzymes have also been shown to prodce proteolytic enzymes.

LIPOLYTIC ENZYMES

Different types of lipids are present in plant cells, the most important of them being the neutral lipids functioning as energy sources, while phospholipids and glycolipids are the main components of cell membranes. Several fungi, bacteria and nematodes are capable of degrading lipids. Plant pathogens produce enzymes that attack major membrane constituents. Lipolytic enzymes, called lipases, phospholipidases, etc. hydrolyse the liberation of fatty acids from the lipid molecule by hydrolysing the acyl ester and phosphate ester bonds of phospholipids. These enzymes and proteases might affect cell membranes since phospholipids and proteins are the major constituents of plant merrbrane systems.

ENZYMES INACTIVATION

In some host and parasite combinations resistance may be related to inactivation or inhibited synthesis of enzymes by the parasite.

In many cases of fungal enzyme inactivation there is proof that the inhibitors are of phenolic origin. Phenols and their oxidation products inhibited a wide range of enzymes

Chona studied that macerating enzymes from a many of plant pathogens are inhibited by juices from apples and potato tubers. This was due to magnesium ions and potassium phosphate in tuber extracts and malic acid in apple juice.

"Byrde" has attributed the resistance of certain varieties of cider apples to brown rot to their high content of polyphenols in parenchyma of the fruit.

In the wilt of tomato, there is inhibition of pectolytic

activity in hyphae invading parenchyma, and the lack of inhibition of mycelium in vessels may account for mycelium being restricted primarily to vessels.

Oxidised phenocls, quinones, and related substances have been employed as chemotherapeutic agents in wilt diseases.

Mahadevan *et al.* (1965) have demonstrated that the extracts of a number of plant species inhibit pectic enzymes *in vitro* and in some instances, occurrence of such inhibitors *in vivo* may be associated with resistance to pathogens.

Colletotrichum lindemuthianllm is capable of making a various polysaccharide-degrading enzymes and the virulence of the isolates has been correlated to their ability to produce α-galactosidase. When the alpha or gamma strain of the fungus is grown in a medium having galactose as the sole carbon source, large quantities of α-galactosidase are secreted. In the presence of either glucose or xylose, α-galactosidase synthesis is decreased β-glucosidase appears to be produced constitutively in the alpha strain of C. lindemuthianum, while β-xylosidase is induced when either the alpha or gamma strain of the pathogen is grown in culture with xylose as the sole source of carbon.

Addition of glucose to the cut stems of *Fusarium oxysporum* f. *lycopersici* infected tomato plants decrease the rate of symptom development and the level of vascular system polygalacturonase activity. Glucose also represses the synthesis of polygalacturonase activity.

Another way in which plants are protected against cell wall degrading enzymes secreted by pathogens is by a modification of the primary cell walls.

The development of resistance in the hypocotyls of maturing Red Kidney beans to infection by *Rhizoctonia solani* is correlated with the conversion of pectin to calcium pectate in the hypocotyl cell walls. Calcium pectate is resistant to the polygalacturonase secreted by the fungus.

Albersheim and Anderson, while describing the cell wall and wall breaking enzymes in the infection process, says that a pathogen is incapable of invading a plant unless the former finds the best environment to permit it to secrete sufficient amounts of endopolygalacturonase to overcome the amount of inhibitor present in the cell walls of the plant. Although there are several endopolygalacturonase inhibitors in the cell walls of dicotyledone, each of them appears to. be able to inhibit the endopolygalacturonases of a variety to unrelated plant pathogens.

The ability of plants to produce enzymes that are able to break down the cell walls of pathogen another general defense mechanism. Plants can produce large amounts of lysozyme, chitinase and endo-β -1,3 glucanase. The chitinase of tomatoes appears to participate in the lysis of Verticillium albo-atrum mycelia that occur during the infection of tomatoes by this pathogen.

TOXINS AND PLANT PATHOLOGY

There is an excessive amount of studies on toxins and their role in plant diseases.

Although, the germ theory of disease for plants was described earlier than humans yet the role of toxins in infections and in diseases was first established in human medicine. The term toxin in plant pathology has actually been adopted from human medicine. The idea that pathogenesis might be due to the production of poisons by the pathogens is by 110 means new, but it is only in recent years that it has been put on a scientific footing.

De Bary showed that plant pathogens generate toxins but experimental evidence to support this theory was available only in 1913 by Hutchinson. The credit for the discovery of toxin is generally given to Roux and Yersin who first saw the presence of a toxin in the disease caused by *Corynebacterium diptheriae.* In the diseased animals he observed symptoms very far from the site of infection. They conceived that this may not be directly due to the microbe itself but to its metabolic product which gets transformed and reveals symptoms at distant places from the actual site of infection.

A toxin can be defined as a product of microbial origin involved in host pathogenesis. Toxins are generally proteins which are antigenic in nature and have antitoxins. Also by given small doses of a toxin it is possible to develop immunity in the host against the toxin, and the producing it.

Plant pathologists have doubts whether there are toxin in phytopathology. Actually, They are not toxins in the real sense, playing their part in the plant disease syndrome, as in human or animal pathology, because none of these satisfy the conditions of a toxin in mammalian pathology.

The concise Oxford Dictionary defines a toxin as, "a poison, especially one secreted by a microbe and causing some particular disease, and a poison as, "a substance when introduced into or absorbed by ad living organism destroys life or injures health, especially one that destroys life by rapid action and when taken in small quantity".

Dimond and Waggoner defined a toxin as a phytotoxic compound produced by a micro-organism. This excludes toxic substance, secreted by plant parasites other than a microorganism. Such restrictions are unnecessary. They also recognized vivotoxin as a substance produced in the infected host by the pathogen and/or its host which functions in the generation of a disease, although it is not itself the initiating agent of the disease.

Toxins are different from enzymes as they do not attack the structural integrity of the tissues but change the metabolism in a subtle manner. It is the subtility of action that differentiates toxins from enzymes. Toxins act directly on the protoplasts of cells and substances with the same end effect but an indirect method of action are not toxins. The toxin of *Clostridium welchii,* lecithinase, even though an enzyme, is a toxin because it break down cell membrances. Such enzymes are toxins.

CLASSIFICATION OF TOXINS

Wheeler and Luke defined three broad classes of toxins, according to their source of origin namely, a) phytotoxins, b) vivotoxins and c) pathotoxins.

Phytotoxin

Any compound secreted by a micro-organism which is toxic to plants is a phytotoxin. Phytotoxins are nonspecific, produce few or none of the symptoms that are produced by the pathogen, and as happens in most cases, show rio relation between toxin production and pathogenicity. Lycomarasmine and alternaric acid are examples of phytotoxins.

Vivotoxins

Vivotoxin can be defined as "as a substance produced in the infected host by the pathogen and or its hosts which functions in the production of the disease but is not itself the initial inciting agent of the disease". Certain criteria which should be fulfilled before being called a vivotoxin are:

A. It must be isolated from the diseased plant but should not be present in the healthy plant.

B. It must be characterized chemically.

C. When introduced in a pure form into a healthy host, it must produce the symptoms of the disease or a portion of the syndrome. Fusaric acid and piricularin are typical examples of vivotoxins.

Pathotoxins

Pathotoxins play a causal role in disease production, and produce symptoms characteristic of the disease in susceptible plants (Wheeler and Lukė, 1963). These toxins may be produced by the pathogen, host or interaction between them. Piricularin produced by *Piricularia oryzae* is a typical pathotoxin.

Fusaric Acid

Fusaric acid (5-n-butyl pyridine-2 carboxylic acid) is a toxin which is produced by several species of *Fusarium*, mainly belonging to the group Elegans Woolenweber it causes wilt diseases in tomato, cotton and other plants. Some of these fungi also produce other toxic metabolites which may be closely related to fusaric acid.

Fusaric acid was, first seperated as the metabolic product of *Fusarium heterosporum* Nees, a non-specific parasite, by Yabuta. In 1952, it was recognized as an important toxin of *Fusarium Iycopersici* Sacc:, *Fusarium vasinfectum* Atk., and *Gibberellafujikuroi.* It is now known to playa pathogenic role in the wilt of tomato and cotton

Fusaric acid is also called, a wilt toxin, as it is plays big role in the wilt of tomato and cotton plants than in any other disease. Some of the toxic symptoms on diseased plants could be produced by pure toxin fusaric acid. Moreover; it has been firmly established that a concentration of 10^{-4} M of fusaric acid is reached in the infected plants, a concentration lethal to the host tissues.

The symptoms of fusaric acid injury are produced first in the stems and soon after in the leaves in the form of intercostal necrosis of leaves. In the stem it is mainly the cortical tissue over the vascular bundles that is destroyed and light grey-green sunken furrows appear on the stems and gradually spread to the petioles. Next the petioles start to bend downwards. This is called epinasty and this injury is similar to that caused by ethylene. With higher concentration, the whole stem tissue is weakened and necrosis of the cortical tissues overlying the vascular bundles in the stem sets in. In the leaves, necrosis extends in the interveinal areas, and subsequently, along the whole length of the veins. About 24 hours after toxin intake the leaves curl and become flaccid.

Fusaric acid has the empirical formula $C_{10}H_{13}O_2$ N and molecular weight 179. It has the following structural formula:

$CH_2-CH_2-CH_2-CH_3$

HOOC

N

Fusaric acid

Pure Cusaric acid has a melting point of 98-100°C. But biogenic preparations often contain traces of dehydrofusaric acid, and therefore, possess a higher melting point (up to 109°C). Fusaric acid is toxic to higher plants and micro-organisms interfering with their metabolism in various ways.

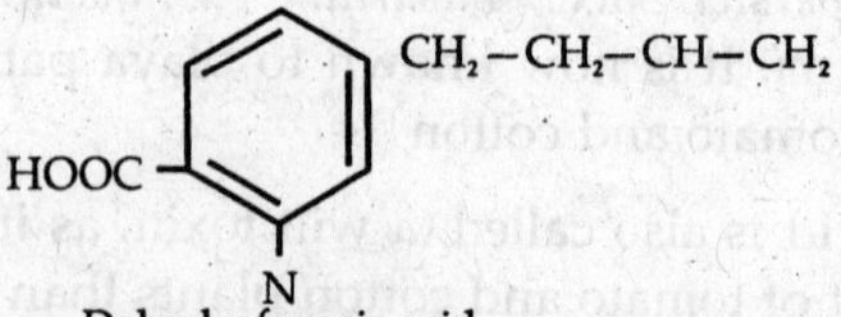

$CH_2-CH_2-CH-CH_2$

HOOC

N

Dehydrofusaric acid

Before the appearance of the symptoms-necrotic spots on the leaf biades, the shrivelling and drying of leaves, and the shrinking and wilting of stem and petioles in tomato cutting—the water permeability of protoplasts and the water balance of the whole plant are altered. Fusaric acid inhibits

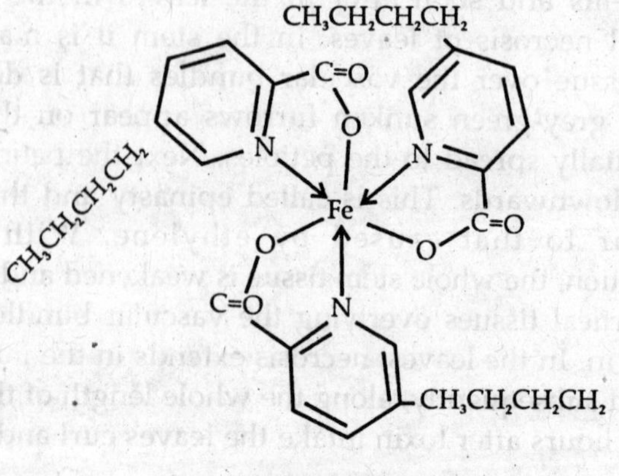

$CH_3CH_2CH_2CH_2$

C=O

O

N

N

$CH_3CH_2CH_2CH_2$

Fe

C=O

O

O

O

C=O

N

$CH_3CH_2CH_2CH_2$

Iron chelate of fusaric acid

polyphenol oxidase competitively in much lower concentrations than other enzymes and may also react with various phenol derivatives.

A comparison of fusaric acid with other pyridine derivatives shows that two parts of the molecule are must for toxicity. Inhipition of fungal spore germination, yeast respiration, and polyphenol oxidase depends on the presence of a carboxyl group in α-position to the nitrogen. The aliphatic side chain in βposition determines primarily the effects on water permeability, the severity of the damage increasing with the length of the side chain methyl to butyl.

Fusaric acid forms stable chelate rings, catching a metal ion between the N-atom of the pyridine ring and carboxyl radical in the α-position, forming 5-membered coordinate link. Of the metal ions, ferric ion is most strongly chelated with it. On the basis of the antagonism reported between ferrioxamine Band fusaric acid, it could be conjectured that fusaric acid specifically interferes with the incorporation of Fe^{+++}in the porphyrin skeleton. The mode of action of fusaric acid has been discussed by Kalyansundram (1970). Sanwal (1961) described the physiological effects of this toxin, and suggested that it could cause a partial uncoupling of oxidative phosphorylation. According to Deverall (1964), it plays no part in the development of wilt symptoms, although it might act in conjunction with other metabolites.

Lycomarasmine

Lycomarasmine is a, wilt toxin of the tomato wilt pathogen *F. Iycopersici* Sacco This toxin, unlike fusaric acid, appears to be specific to F. Iycopersici. Lycomarasmine, when administered in tomato plants, produces injury in the intercostal fields of the leaves, and rolling of the leaf tips. The symptoms of tomato wilt look nice the toxic injury produced by the synergistic action of Iycomarasmine and fusaric acid. The minimum dose of lycomarasmine that is needed to cause appreciable injury to tomato shoots is 150mg per kg fresh weight

Lycomarasmine is a polypeptide with molecular weight 277.3 and empirical formula $C_9H_{15}O_7N_3$.

```
H2N-CO-CH2                    CH3
      |                        |
HOOC-CH-NH-CO-CH2-NH-C-OH
                               |
                              COOH
```

However, using the powerful technique of NMR

spectroscopy in a comparative study of lycomarasmine with aspartic acid. Currently accepted as the correct formula for lycomarasmine, is given below

```
HOOC  H
    \ /             COOH
     C–NH–CH₂–C–H
                    |
                    N–H
                    |
                    CH₂
                    |
                 O=C–NH₂
```

Lycomarasmine

The toxicity of Lycomarasmine arise from the fact that it can form metal chelates. Thus, the injury caused by lycomarasmine when administered to tomato plants causes a shock phase within the first few hours when there is a depressed rate of water intake as well as transpiration. Soon this get reversed, and the rate of transpiration exceeds the normal rate, bringing about the wilt. The toxin effect of this toxin is partly due to the fact that its molecules, on their way through the tomato shoots, chelate the iron ions of the host, which would not be able to migrate themselves, and transport them into the leaves in the form of a lycomarasmine-iron complex. Here, part of the complex is decomposed due to photochemical effects or due to exchange by more potent chelating agents in the cell itself. Thus, on one hand, part of the lycomarasmine molecules are regenerated as chelating agents and cause appropriate local deficiencies, while on the other hand, iron ions set free in the 'wrong' place lead to iron plethora.

Another aspect of the study of this toxin has been its anmity to the yeast growth factor, strepogenin which is an oligopeptide containing glutamic acid. As against, glutamic acid which is present in strepogenin, lycomarasmine has aspartic acid, and due to this has its antagonistic action. This growth factor, when given externally, revest some of the toxic symptoms of the lycomarasmine injury in tomato, leading to the conclusion that it may interfere with the strepogenin metabolism of the host. The role of toxin lycomarasmine in the development of disease symptoms has been disputed by a number of workers.

Piricularin

Pyricularia oryzae is a part of a group of Hyphomycetes, is the causal organism of the blast disease of rice and has a wide distribution in all rice growing areas of the world. It appears that two of the products of the fungus are responsible for the characteristic symptoms of the disease. One of them is identified as α-picolinic acid and the other is piricularin. Piricularin is more active and semispecific in that it affects blast susceptible rice plants more than resistant ones. Low concentrations of piricularin stops the germination of the conidia of *P. oryzae,* and the fungus has been shown to produce a piricularin binding protein, a copper oxidase which binds piricularin and destroys its fungitoxicity but not its phytotoxicity. The ways in which piricularin affects plant tissues ate not clear but it brings aboutan an increase in polyphenols and oxidases and its toxicity is countracted by chlorogenic acid, one of the principal polyphenols of the rice plant. It increases respiration and growth at low concentrations but inhibits them at higher ones.

Piricularin has been assigned a tentative empirical formula of $C_{18}H_{14}N_2O_3$, melting point of 73.5°C a nd a molecular weight of 306. It is highly stable in water, especially at low *pH*.

The toxic action of piricularin appears to be due to its negative effect on the activity of several enzymes such as peroxidase, catalase, cytochrome oxidase and ascorbic acid oxidase, even at very low dilutions (1/100,000-1/200,000). At still lower concentrations (1/1,600,000), the toxin stimulates the respiration and growth of rice plants.

The toxic effect of piricularin is made neutral by chlorogenic acid or ferulic acid, each of them being the natural phenolic constituent of the rice plants. It is presumed that the combination of piricularin with these substances is a natural detoxifying mechanism.

A third substance, pyriculol ($C_{14}H_{16}O_4$) was isolated from cultures of *P. oryzae* which produces natural blast lesion and also inhibits the growth of rice seedlings. The compound is a heptaketide

In addition to the above three toxins, two more toxic substances have been recorded by Iwasaki, namely, 3,4 dihydro 3,4,8 trihydroxy-l (2H) naphthelenone and Tenuazonic acid. The latter substance has been isolated from infected tissues also, and thus, is a vivotoxin.

Victorin

In the late 1940s, oat crops in the USA were devastated by a seedling blight caused by *Helminthosporium victoria* in the variety victoria or its derivatives. Although the pathogen was typically localized in the basal portion of infected plants, symptoms appeared on the leaves which often collapsed. Litzenberger demonstrated that the culture filtrates of the fungus contained a toxin that can reproduce accurately the visible symptoms of the disease. What was particularly interesting about this toxin was that, resistant oat cultivars were not affected by it. The name "victorin" was given to this toxin by Wheeler and Luke. The toxin is specific in its activity and is claimed to produce the symptoms only in the pathogen-susceptible variety of oats. Number of toxins with host specificity and an ability to incite all of the symptoms associated with a particular disease have been identified. These are known as host-specific toxins to distinguish them from those which are nonspecific.

Victorin is by far the most potent and selective pathotoxin known on a dry weight basis; retined preparations produce toxic effects on oat plants susceptible to *H.victoriae* at concentration below 2×10^{-4} *pg*/ml.

Victorin which can be substituted *for H.victoriae,* has made it possible to follow metabolic change in diseased plants without complications by the metabolic activities of a living pathogen. Since changes produced by victorin are almost similar to those which occur in plants infected by *H. victoriae* or many other pathogens, victorin has provided a important tool for investigating the nature and sequence of events which occur during pathogenesis. The earliest effect detected was in permeability. Victorin causes a decrease in the C_6/C_1 ratio of susceptible oat leaves. This gives additional evidence that victorin induces metabolic alterations similar to those found in other diseased plants.

The chemistry of this toxin is not yet completely deciphered, but current studies shows are that it is a polypeptide linked to a tricyclic secondary amine with a molecular weight of around 2000. When pure toxin is treated with saturated sodium bicarbonate for 24 hours at room temperature, it breaks into two substances. One of the products is a peptide containing aspartic acid, glutamic acid, glycine, valine and one of the leucines. The second product of sodium bicarbonate cleavage was crystallized and was found to be a new base called Victoxinine. This has an empirical formula of $C_{17}H_{29}NO$. The evidence indicates that it is a tricyclic secondary amine. Victoxinine is less toxic than the parent molecule.

Victorin is able to produce toxic symptoms only in the blight susceptible variety of oats. Perhaps, this could be attributed to the presence of a specific enzyme on the surface of the cells of the susceptible and not the resistant variety. Recent studies shows that the initial effect of victorin is not the disruption of plasmalemma but is on the cell wall or some readily accessible surface component and protoplast permeability is disrupted only after a lag of 30 minutes.

There are many other host specific toxins, like *Helminthosporium carbonum* toxin or HC toxin, *Helminthosporium sacchari* toxin or HS toxin, *Helminthosporium maydis* toxin or HM-T toxin, *Periconia circinata* toxin or Periconin, and *Alternaria kiktichiana* toxin which are host-specific.

Wildfire Toxin

Pseudomonas tabaci produces a phytotoxin called *wildfire toxin,* this name being given from a description of the disease on tobacco. Recent studies strongly suggests that *P. coromofaciens,* which attacks oats, and an undescribed pseudomonad which attacks timothy also synthesizes the same compound. Braun proposed that the toxin interferes with methionine utilization. For many years this toxin produced by *P. tabaci* was thought to be an antimetabolite of metbionine. The main defect with this hypotbesis was, the failure of methionine to protect tobacco plants against the effects of the toxin or the pathogen. More recent studies indicates that the toxin inhibits glutamine synthetase (GS) activity and that tobacco leaves infiltrated with glutamine are protected from the effects of the toxin.

Lovrekovich, suggested that the wild fire toxin may directly affect RNA metabolism. In diseased leaves, the concentration of RNA and soluble proteins decreased concomitantly with increases in free amino acids and ammonia, an effect counteracted by the addition of kinetin. This toxin does not seem to affect protein synthesis directly.

Woolley showed that the wild fire toxin made up of the amino acid tabtoxinine (α-,ϵ- diamino-β-hydroxypimelic acid) attached to lactic acid by a lactone ring; subsequent work disproved the structure.

Based upon extensive NMR experiments, along with basic compositional data and other spectroscopic observations, Stewart presented what is the correct structural formula of tabtoxin.

```
                                          NH2
                                          |
       H   O   H   H                   H–C–COOH
       |   ‖   |   |                      |
 H2N–C – C – N – C–COOH                  CH2
       |           |                      |
      CH2       H–C–OH                   CH2
       |           |                      |
  O   CH2         CH3  ──────→       H2N–C–COOH + Threonine
  ‖    |                                  |
  C – C   OH                             CH
  |    |                                  |
H–N – CH2                                OH

Tabtoxin                             Tabtoxinine
```

Hydrolysis of tabtoxin yields tabtoxinine and threonine

Similar results were found by Taylor. From these data it has been suggested that the bacteria synthesize two closely related compounds, one containing serine and the other threonine. Whether only one or both are phytotoxic is unknown at present. Crude toxic preparations from *P. tabaci* have been called the wild fire toxin. Since at least two toxins have been isolated from such preparations, the name 'wild fire toxin' no longer seems appropriate. Patil (1974) has discussed this aspect in detail.

Alternaric Acid

Alternaric acid occurs in the early blight of tomato and potato-and is caused by *Alternaria solani.* Some of the lesions appear to be devoid of the fungus under certain conditions, suggesting that a mobile toxin is involved. Cultures of *A. solani* produced a dibasic acid alternaric acid-which caused necrotic lesions similar to those caused by the fungus. The toxin also stops or severely restricts growth *of* the fungal germ tubes without in reality preventing germination. Alternaric acid is fairly nonspecific and its significance in early blight is uncertain. Thus, some highly pathogenic strains *of* the fungus failed *to* produce it in culture and, it is clear that other factors and possibly other toxins are involved.

Alternaric acid produced by *A. 'solani* Jones and *Grout* has antifungal but not antibacterial activities. Also it is highly fungitoxic. This substance is an optically inactive, unsaturated diabasic acid with an empirical formula of $C_{21}H_{30}O_8$. The effect *of* altemaric acid on the water process of host plants is same to that observed with the toxin lycomarasmine but it is highly effective. This substance or a substance closely related to it has been detected in the naturally infected plants and there is considerable proof to show that it is involved in the disease caused by *A. solani.* Although a similarsubstance Was found in an infected tomato fruit, the importance of alteronaric acid as a 0 vivotoxin has not been stressed because of a lack of correlation between the ability of different isolates of *A. solani* to produce the toxins and *to* cause the disease.

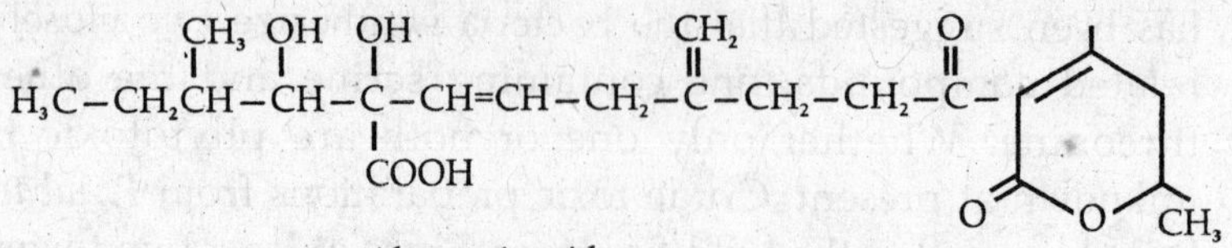

$$H_3C-CH_2CH(CH_3)-CH(OH)-C(OH)(COOH)-CH{=}CH-CH_2-C({=}CH_2)-CH_2-CH_2-C({=}O)-$$ (attached to a 6-methyl-dihydropyran-2-one ring)

Alternoric acid

Alternaria Kikuchiana Toxin

The wide use *of* new crop varieties often produces new disease problems. Thus, the black spot disease of Japanese pears appeared after the new natural mutant variety *Nijisseiki'*was introduced. The casul organism was identified in 1917 as a species *of Alternaria* later named as *A. kikuchiana.*

The isolation procedure involved in extraction of A. Kikuchianua toxin with various solvents, precipitations, and chromatography on alumina. These similar products 0 named phytoalternarin, A, Band C had a high degree *of* selective toxicity. The effect *of* 0, AK toxin on pear appears to be similar to the effect *of* HV toxin on oats. AK toxin caused an increase in electrolyte loss from susceptible but not resistant leaves within 20 minutes *of* initial exposure .

There are several other toxins which have been reported during the last few decades. Some *of* these are host specific while others are non-specific.

7

Nutritional Studies of Plant Pathogens

GROWTH AND DIFFERENTIATION

According to Thompson (1948), "growth is somewhat a vague word for a very complex matter". Needham (1942) has analyzed this complex phenomenon in detail, particularly with reference to the relation between growth and differentiation. While Needham's comments are concerned primarily with the development of animal embryo, they are pertinent to fungi and can help clarify our thinking about the problem. Growth can involve an increase in the number of nuclei, in the number of cells, in the size of the cells or in the amount of 'nonliving' structural material. Hawker and Linton (1971) defined growth as an increase in mass or size accompanied by the synthesis of macromolecules, leading to the production of new organized structure. The kinetics of fungal growth have been discussed by Mandels (I965). Robertson (I968) and Jerebzoff (1965) have also discussed the growth systems in fungi.

Fungi are of very diverse morphological forms, varying from unicellular yeasts to large fleshy basidiomycetes. The morphology of many fungi is markedly dependent upon growth conditions. Some yeasts can be induced to become filamentous and on the other hand, some filamentous fungi grow in unicellular fashion (Cochrane, 1958). Fungi are fundamentally filamentous in organization and their morphology is determined by complex structures, division of an individual cell or differentiation of a hyphal strand. When we are considering the kinetics of fungal growth, we are concerned primarily with two situations-growth as a population of individual cell or growth by the elongation of a hypha.

The precise definition of the growth of fungi depends on the method of measurement used. Most methods determine directly or indirectly the increase in mass of inoculum after a known incubation in a complete medium. Increase in mass or number of cells may be used as an operational definition of growth.

METHODS OF GROWTH MEASUREMENTS

To measure rates of growth, we can take any quantity which is known to be proportional to the criteria for growth being used. The measurement should also be one which is convenient and which has a degree of precision consistent with experimental variables. Growth can be measured by the changes in the number of cells, in linear dimensions, in mass, in volume, in total activity of any metabolic process or in quantity of some cellular constituents.

Linear Measurement

The widely used method of measuring fungal growth consists in growing fungi in petri-dishes and measuring either the diameter or the area of the colony. The increase in colony diameter or radius is recorded at suitable intervals. For fungi which grow very rapidly, or for some special cases, it may be more convenient or useful to employ 'race' tubes (Ryan *et al.*, 1943) or some modification of the same. Linear measurement can also be made on a microscopic scale by observing individual hypha.

The advantages of linear measurements are in their simplicity, and non destructive nature. Repeated measurements can be made on the same culture" thus reducing the number of replications required. There is a fundamental disadvantage in the use of linear measurement for quantifying growth rates; there may not be any correlation between the spread of mycelial front on a solid surface and the total amount of fungus produced.

Dry Weight Method

This method is probably the most widely used and the most generally applicable for assaying growth in fungi. By weighing the mycelium and spores produced, an accurate and objective measure of growth is obtained. Although dry weight is a basic quantity which can be easily measured, there are certain limitations and disadvantages in its use. It is a destructive method and consequently requires a large number of cultures to provide a significant measurement of the course of growth. It is most widely and usefully used in liquid cultures which are shaken to provide a more uniform environment and better gas exchange. However, this is not the normal environment for most organisms.

Cell Volume Method

Cell volume determination can be convenient and useful criterion in certain specialized cases where growth occurs in a unicellular manner, such as in yeasts and bacteria or in studies involving spore germination (Mandels and Darby, 1953). This method is not useful where filamentous growth occurs because of the nonuniform packing. However, it has certain advantages. It is simple, quick and needs only small quantities of the materials required. The techniques include centrifugation in hematocrit-tube or the use of a device such as the Coulter counter (Kubitschek, 1960).

Cell Number Method

This is applicable in those organisms which are unicellular, such as yeasts. In addition to yeast certain normally filamentous fungi may grow as unicellular organisms and the technique employed by the bacteriologist, is applicable. Changes in light absorption, scattering or devices such as the Coulter counter and the de Bonet Maury Bio-spectrophoto meter are used.

Metabolic Activity

Growth can also be measured indirectly by following the metabolic activities of the organism, such as measurement of respiratory activity (O_2 consumption or CO_2 liberation), product formation (acids, pigments, etc.) and disappearance of substrates of enzymatic activity. The advantages of this method are that it is simple, non destructive and continuous measurements can be made. This method is of particular value when transitory changes occur, such as in germination or in adaptation to new growing conditions. An O_2 analyzer or infrared CO_2 analyzer coupled with continuous recorders may be of use in finding out the extent of growth.

Compositional Alteration

Growth can also be determined by following the changes in the quantity of some cellular constituents or the total organic matter (Lu *et al.*, 1959). The determination of growth by cellular nitrogen can give an idea of growth.

KINETICS ANALYSIS

The manifestations of growth in fungi are so varied that no generalized analysis can be made. In certain cases, however, useful mathematical expressions have been shown to be applicable. These situations are discussed below.

A diagramatic growth curve showing an increase in the number of cells is given in Figure. The data are also plotted on a semi log basis to demonstrate the exponential nature of the curve. It is evident that the growth can be divided into several phases.

a. *Lag Phase.-In* this phase, no cell division occurs but cell enlargement and an increase in metabolic activity may occur. This phase may be long or short depending upon the age, size of the inoculum, vigour of the mycelium, the medium and other factors.

b. *Acceleration Phase-In* this phase, there is an increase in cell division from zero to some maximum value characteristic of the next or exponential phase.

c. *Exponential Phase-* This is also called logarithmic phase. In this phase, propagation takes place at a uniform rate of cell division. This phase is clearly defined for bacteria and approached by yeasts. It is characterized by constant generation time. If the logarithms of the cell number are plotted against time, the curve is a straight line.

d. *Deceleration Phase – This* phase is characterized by a decreasing rate of cell division. As the nutrient becomes exhausted, toxic metabolites may accumulate. These materials, earlier called staling substances, obviously retard growth. If fresh medium is provided and the toxic metabolites are removed, there is delay in the decline of this phase.

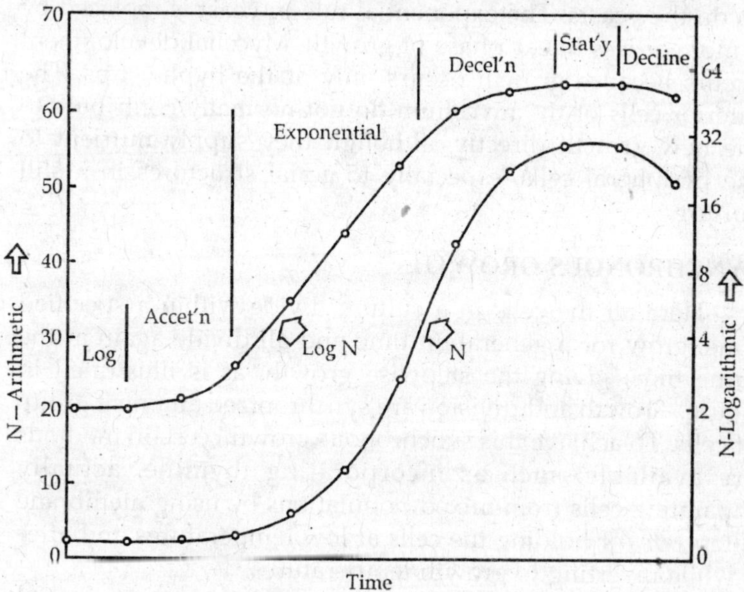

Figure. Idealized representation of exponential growth and phases of growth curve.

e. *Stationary Phase-In* this phase, there is cessation of cell divisions. In fact, there may be a condition when cell division is balanced by death of cells. The interpretation depends upon the techniques used to measure the cell numbers, that is, whether the total cells or viable cells are counted.

f. *Phase of Decline or Autolysis-In* this phase, there is a decrease in the number *of* viable cells. Sooner or later, following maximum development, autolysis sets in. As the cells appear vacuolated cellular enzymes begin to digest various cell constituents. Only the more resistant portion of the cell remains. A microscopic examination at this stage reveals that many cells are devoid of protoplast. It is just possible that some of the material released by the autolysis is used by the remaining cells.

With the exception of the exponential phase of growth, the filamentous fungi follow the same order of development as do the yeasts. The exponential rate in yeast is replaced by a more or less linear phase of growth. Mycelial development occurs because growth occurs only at the hyphal tips. The interior cells of the mycelium do not normally contribute to the new growth directly, although they supply nutrient to the peripheral cells, especially to aerial structures in a still culture.

SYNCHRONOUS GROWTH

Here all the cells in a culture divide within a specified time, grow for a generation time and all divide again at the same time, giving the stepwise growth, as is illustrated in figure. Growth and division are synchronized among a group of cells. To achieve this synchronous growth, certain methods are available, such as incorporating thymine, actually separating cells from mixed populations by using membrane filters, etc. or holding the cells at low temperatures and after a while reverting to growth temperatures.

RHYTHMS IN GROWTH

Jerebzoff (1965) has described the various aspects of growth rhythms in fungi. Many fungi exhibit rhythmic changes in their growth and activities, which may either be due to periodic fluctuation on external physical conditions (exogenous rhythms) or to the inherent characteristics of the organisms. Formation of zonation, that is, growth in concentric circles is the most common rhythmic expression manifested on solid media or on host tissue.

FACTORS EFFECTING GROWTH

In addition to nutrition, there are other factors which can affect growth. There are internal factors which are dependent on the genetic constitution of the organism as well as external factors, such as temperature, pH, light, moisture, O_2 and CO_2 concentration, etc. Lilly and Barnett (1951), Cochrane (1958) and Bilgrami and Verma (1978) have discussed these factors in detail.

CULTURE MEDIA

Fungi are dependent on the medium or the substrate for all the elements and compounds they require or utilize, except molecular oxygen and possibly a little carbondioxide which are obtained from the atmosphere. Nutritional requirements of various fungi differ and there is no one medium or substrate which can be universally suited to all the fungi. The genetic constitution of the fungus determines what it can do, but the expression of its potentialities is dependent upon the composition of the medium on which it grows and the environment to which it is exposed. The choice of the media and environmental condition depend on the purpose of the investigator. Micheli (1679-1737) was the first botanist who cultured fungi on pieces of watermelons. The use of such substrate for the artificial culture of fungi continued till Pasteur (1860) used what might be considered a proximation of a chemically defined medium during his studies on alcoholic fermentation. Credit goes to Raulin (1869), a student of Pasteur, who first devised a synthetic medium for the nutritional studies of some common fungi. Raulin's medium had the following constituents:

Ammonium nitrate	4.00 g
Ammonium phosphate	0.06 g
Magnesium carbonate	0.04 g
Potassium carbonate	0.06 g
Ammonium sulphate	0.25 g

Zinc sulphate	0.07 g
Iron sulphate	0.07 g
Potassium silicate	0.07 g
Sucrose	70.00 g
Tartaric acid	4.00 g
Water	1500.00 ml

However, inadequate information was given for the duplication of this medium. It is still not known which ammonium phosphate $(NH_4) H_2PO_4$ or $(NH_4)_2HPO_4$, which zinc sulphate $ZnSO_4 . 7H_2O$ or $ZnSO_4' H_2O$, which iron sulphate, $FeSO_4 . 7H_2O$ or $Fe_2 (SO_4)_3 7H_2O$, which tartaric acid, D-tartarate or L-tartarate or D-tartarate or meso-tartaric acid were used by Raulin in the original work.

Brefeld (I881) gave the following directions for preparing a synthetic medium. He suggested adding cigar ashes, dissolved in nitric acid or citric acid to a solution containing a soluble carbohydrate, such as glucose and an ammonium salt. The amount of ashes was not specified and this cannot be said to be a synthetic medium.

Naming culture media after personalities may be of historical interest but this practice is unfortunate and should be avoided. Lilly and Barnett (1951), among others, suggested that we should designate the media by naming the carbon and nitrogen sources. Accordingly, Czapek's and Schopfer's medium should be designated as sucrose nitrate and glucose asparagine media, respectively.

CLASSIFICATION OF CULTURE MEDIA

Media can be classified on the basis of chemical composition, physical state and utility purpose. According to the use it is put to the medium may be categorized as follows:

a. Routine laboratory media, such as PDA, Czapek-Dox Agar, Malt Extract, etc.

b. Enriched media, such as Peptone Martite Agar.

c. Selective media which includes such as medium for the isolation of *Fusarium* spp. and Pimaricin Vancomycin, and Pentachloronitro-benzene-medium for the isolation of pythiaceous fungi.

d. Differential media which aid in differentiating between various kinds of organisms, such as *Botrytis* separation agar used to distinguish between *B. cinerea* and *B. alli;* the latter species is sharply restricted by sorbose.

e. Assay media-Media employing for assaying amino acids and antibiotics.

f. Biochemical media-These media are generally used for the differentiation of micro organisms on the basis of their biochemical activity.

Media can also be classified on the basis of their chemical compositions. These are classified into natural, semi-synthetic and synthetic media. The chemical composition of a natural medium is not properly known. In a semi-synthetic medium, the chemical composition is partially known and in a synthetic medium, the chemical composition is completely known. The following is the partial list of some of these media:

Natural Media

Various fruits and vegetables, raisins, potato plugs, carrot slices, twigs, stems, roots and leaves of various plants.

Semi-synthetic Media

Any medium which contains agar becomes a semi-synthetic medium. Czapek-Dox agar, loatmealagar, potato dextrose agar, beef peptone or nutrient agar are some of the semi-synthetic media.

Synthetic Media

Some of the commonly used synthetic media are Czapek solution, Raulin's solution, Richard's solution and Martin's rose bengal medium, etc.

The following is a list of some of these culture media (with their constituents) which are useful in plant pathological and microbiological work:

CULTURE MEDIA

A. GENERAL, PURPOSE MEDIA

(a) Cornmeal Agar (Riker and Riker; 1936)

Agar	17.0 g
Cornmeal	20.0 g
Peptone (if desired)	20.0 g
Dextrose (if desired)	20.0 g
Water (distilled)	1000.0 ml

Use: A common medium for the cultivation of fungi.

(b) Czapek-Dox Agar (Raper and Thorn, 1949)

Agar	15.0 g
$NaNO_3$	2.0 g
K_2HPO_4	1.0 g
$MgSO_4.7H_2O$	0.5 g
KCl	0.5 g
$FeSO_4, .7H_2O$	10.0 g
Sucrose	30.0 g
Water (distilled)	1000.0 ml

Use: A common medium for the cultivation of fungi and bacteria.

(c) Malt Extract Agar (Raper and Thorn, 1949)

Agar	25.0 g
Malt extract (Difco)	20.0 g
Dextrose	20.0 g
Peptone	1.0 g
Water (distilled)	1000.0 ml

Use: A common medium for the cultivation of fungi.

(d) Nutrient Agar (Difco Manual, 1953)

Agar	15.0 g
Beef extract	3.0 g
Peptone	5.0 g
NaCI	8.0g

Water (distilled)	1000.0 ml
pH	7.3

Use: For cultivation of bacteria not requiring a high nutritious medium.

(e) Oatmeal Agar (Johnson and Curl, 1972)

Agar	17.0 g
Oatmeal	15.0 g
Yeast extract (if desired)	1.0 g
Water (distilled)	1000.0 ml

Use: A common medium often used for maintaining stock cultures of actinomycetes and fungi.

(f) *Peptone Dextrose-Rose Bengal Agar* (Martin, 1950)

Agar	20.0 g
KH_2PO_4	1.0 g
$MgSO_4.7H_2O$	0.5 g
Peptone	5.0 g
Dextrose	10.0 g
Rose Bengal (1 %)	3.3 ml
Water (distilled)	1000.0 ml
Streptomycin	30.0 mg

Use: A general purpose medium for isolating fungi from soil.

(g) Potato-Dextrose Agar (Riker and Riker, 1936)

Agar	17.0 g
Potato (peeled and sliced)	200.0 g
Dextrose	20.0 g
Water	1000.0 ml
pH	6.0-6.5

Use: A general purpose growth medium for many fungi. *(h) Soil Extract Agar* (Allen, 1957)

Agar	15.0 g
Glucose	1.0 g

K_2HPO_4	0.5 g
Soil Extract	100.0 ml
Water	900.0 ml
pH	6.8-7.0

Steam 1000 g garden soil in 1000 ml of water in an autoclave for 30 minutes at 1.05 kg cm^2 pressure. About 0.5 g of $CaCO_3$ is added and soil suspension is filtered through double filter papers until it is clear. The filtrate is sterilized and stocked for further use.

Use: A general purpose medium for isolating bacteria from soil.

(i) *Starch-Casein Agar* (Kuster and Williams, 1964)

Agar	18.00 g
Statch	10.00 g
Casein (vitamin free)	0.30 g
KNO_3	2.00 g
NaCI	2.00 g
K_2HPO_4	2.00 g
$MgSO_4 . 7H_2O$	0.05 g
$CaCO_3$	0.02 g
$FeSO_4.7H_2O$	0.01 g
Water (distilled)	1000.00 ml
pH	7.0-7.2

Nystatin and Actidione (50, mg/ml of each) should be added *to* the cooled liquid medium before pouring into dishes.

Use: A general purpose medium for isolating actinomycetes from soil.

2. SELECTIVE MEDIA

(a) *BNPRA Medium* (Masago *et al.*, 1977)

Agar	17.0 g
Potato (peeled and sliced)	200.0 g
Dextrose	20.0 g
Water	1000.0 ml

3-hydroxy-5-methyl isoxazole (HMI)	50.0 mg
Benomyl	10.0 mg
Nystatin	25.0 mg
Rifampicin	10.0 mg
Ampicillin	500.0 mg

Use: For direct isolation of *Phytophthora* spp. from soil and plants.

(b) *Mali Extract Agar*

Malt extract	25.0 g
Agar	15.0 g
Water	1000.0 ml

Use: For growth of wood-destroying and many other fungi.

(c) *PCNB Agar, Modified* (Papavizas, 1967)

Agar	20.0 g
Peptone	5.0 g
KH_2PO_4	1.0 g
$MgSO_4.7H_2O$	0.5 g
Water (distilled)	1000.0 ml
Oxgall	1.0 g
PCNB	0.5 g
Chlorotetracyclene HCI	50.0 mg
Streptomycin sulphate	100.0 mg

Use: For isolating fusarium from soil.

(d) P_{10}VP Medium (Tsao and Gilberto, 1969)

Agar	20.0 g
Potato (peeled and sliced)	200.0 g
Dextrose	20.0 g
Water (distilled)	1000.0 ml
Pimaricin	10.0 ppm
Vancomycin	200.0 ppm
PCNB	100.0 ppm

Use: For direct isolation of *Phytophthora* spp. from soil and plant.

(e) *Rice Polish Agar*

Rice Polish	20.0 g
Agar	17.0 g
Water (distilled)	1000.0 ml

Use: For sporulation of *Piricularia oryzae.*

(f) *Rose Bengal Agar, Modified* (Tsao, 1964)

Agar	17.0 g
Dextrose	10.0 g
Yeast extract	0.50 g
KH_2PO_4	0.50 g
$MgSO_4.7H_2O$	0.50 g
Peptone	0.50 g
Rose Bengal	0.05 g
Water (distilled)	1000.0 ml
Streptomycin	0.03 g

Use: For isolating *Thielaviopsis basicola* from soil.

(g) *Sach's Agar* (Luttrell, 1958)

Agar	20.0 g
$Ca(NO_3)_2$	1.0 g
K_2HPO_4	0.25 g
$MgSO_4$ '$7H_2O$	0.25 g
$FeCl_3$	Traces
$CaCO_3$	4.0 g
Water (distilled)	1000.0 ml

Use: Extensively used with corn and kernels or leaves to induce perithecia of *Helminthosporium* spp.

(h) *Soybean Meal Agar* (Smoot *et al.*, 1958)

Ground soybeans	15.0 g
Agar	20.0 g
Water	1000.0 ml

Use: To obtain oospores of *Phytophthora infestans* grown in the dark at 20°C.

Media can also be classified on the basis of physical state, such as solid media, liquidmedia or semi-solid media.

METHODS OF STERILIZING

Autoclaving under pressure is a rapid and efficient method of sterilizing the medium. However, in certain cases, autoclaving may change the constituents of the medium and it may be necessary steam the latter for three successive days in Arnold to Steam Sterilizer. The medium may also be filtered through a millipore filter, sintered glass filter, Seitz filter (consisting of an asbestos disc) or Berkefeld filter (made of fossil diatomaceous-earth). Alternatively, the media may be sterilized by chemical agents, such as ethylene oxide or propylene oxide or by radiation. Ultraviolet radiation of less than 330 μ or high energy radiations including X-rays or γ-rays and high speed electrons, may be used to kill micro-organisms in high doses, but these are not generally used in routine experimentation because of the capital outlay and the inconvenience of the operation.

NUTRITION OF FUNGAL PATHOGENS

Nutrition refers to the food requirements of a living system; it is the sum of the process by which an animal or plant absorbs or takes in and utilizes food substances. While the basic nutritional elements required by most organisms are similar, the quantity needed varies as does their capacity to utilize different sources of inorganic and organic substances to obtain the required elements. Among the living systems on earth, micro-organisms possess the most remarkable capacity to utilize various organic and inorganic substances.

The study of the nutrition of plant pathogens is of paramount importance. Unless plant tissues supply, all the nutrients necessary for the growth of the pathogens, it will not be able to colonize successfully. So fat as phytopathogenic organisms are concerned, although their nutritional needs are broadly similar, they maintain their individuality in their choice of specific substances. Bilgrami and Dube (1976) have summed up this behaviour as one showing fundamental unity with maximum diversity. Phytopathogens are continuously struggling for their survival as they have to procure their nutritional quota from the host. In their saprophytic phase, pathogens have to obtain their nutritional requirements from various substrates in the soil. The capacity to utilize the various substrates for the growth of the phytopathogens is dependent upon the various enzyme systems of the pathogen by which it obtains food. Therefore, not only are nutritional studies of plant pathogens essential in the laboratory, but related studies on the nutritional status of the host also become relevant. A knowledge of the nutrition of plant pathogens is not only of academic interest but is also helpful in certain cases for finding out specific control measures. The nutritional status of the host can be varied to the disadvantage of the pathogen. The nutritional aspect of host parasite relationship has been discussed by Garber (1954, 1956). Garber (1954), working with biochemical mutants of *Erwinia aroideae* which required amino acids, found the virulence of specific mutants related to the availability of the required nutrient at the tissue surface. Garber (1955) reported that a methionine-requiring mutant of *Agrobacterium tumefaciens* was pathogenic on *Bryophyllum,* whereas a mutant which required tryptophan was not pathogenic and one which required both methionine and adenine showed variable pathogenicity.

There are reports that resistance or susceptibility is related to nutritional factors. Lukezic and DeVay (1964) have correlated the resistance of plum varieties to *Rhodosticta quercina* with the amount of myo-inositol in the bark of the tree. The high sugar content of tissues of certain hosts favours some pathogens, such as rusts and powdery mildews, while in the case of *Alternaria solani* and *Ceratostomella ulmi* (Dutch elm disease) the reverse holds good.

MINERAL REQUIREMENTS OF PATHOGENIC FUNGI

Fungi need about 17 elements to meet their nutritional requirements. These are known as essential elements. These elements are utilized in the form of specific compounds as ions and free elements. Some of the essential elements are required by all fungi. Other elements are required only by certain fungi. The essential elements are C, H, O, N, K, P, Ca, S, Mg, Fe, Mn, Zn, Cu, Mo, Sc, V, Ga. The last,three are needed by a few fungi.

Biologically Essential Elements

The mere presence of an element in the fungal cell does not necessarily mean that it is an essential element This element is indispensable in that no other element can replace it. Without these elements, life is impossible. An essential element needed in an extremely small amount may be just as essential as carbon, which comprises almost half the weight of the fungus. The essentiality of an element can be determined by a chemical analysis of the mycelium and spores.

On an average, about 70% of the weight of fresh mycelium is water, while a spore contains about 40% water. An ultimate analysis of mycelium and spores always reveals the presence of carbon and nitrogen. On an average, about 40% dry mycelium is carbon. The percentage of nitrogen found is variable. P, K, Mg, Ca, Na, S, Fe are found in the ash obtained after burning the spores and the mycelium.

Some other elements, such as Cr, Bi, Pb, and Au may be present but they may not necesarily be essential: Elements in fungal ash occur only in minor traces and in such cases it is desirable to approach the problem of essentiality in another way. This is done by omitting the element in question from the medium.

FUNCTIONS OF ESSENTIAL ELEMENTS

Thatcher (1934) has classified the essential elements into three groups depending on their functions: (a) structural, (b) functional and (c) those utilized in the transfer of energy.

a. Structural Elements

Elements utilized in the structural make-up of the organism are generally non-metallic. They are C, N, H, O, S,

P. These elements are also important functionally.

b. Functional Elements

The essential metallic elements may be classified as functional elements. They may also have structural properties, such as Na, Mg, Fe, Zn, Mn, Mo, etc.

c. Elements Utilized in Energy Transfer

These elements are utilized in energy transfer. Phosphorus is such an element.

The essential metallic elements function in conjuction with the enzyme system. This accounts for the small amounts of the elements required. The ratio as well as the amounts of the various essential elements and ions affect certain metabolic activities other than growth. The absolute amounts of the essential elements required differ widely.

The essential metallic elements are divided into two groups-macro and micro-elements on the basis of the amounts ordinarily applied in culturing fungi under laboratory conditions.

Macro-elements needed for fungal growth are C, H, O, P, K, N, S, Fe and Mg. Micro-elements are needed in very small amounts. Fe. Zn, Mn, Cu, Ca, Ga, V, Sc, Mo, Co, are the micro-elements needed for fungal growth.

Potassium (K)

Among all the metallic minerals, potassium is present in the largest amount in both mycelium and spores. Concentrations of 0.001 to 0.004 M of this metal are adequate for most of the fungi (Steinberg, 1946; Jarvis and Johnson, 1950). The essentiality and physiological role of potassium in fungal metabolism have been little investigated. The most striking effect of potassium deficiency in *Aspergillus* niger is an increase in oxalic acid accumulation. A very low supply *of* this element is accompanied by poor sugar utilization, and it seems likely from work with organisms other than fungi that potassium has an essential function in carbohydrate metabolism.

Iron (Fe)

Iron seems to occupy an intermediate position between the macro- and the micro-essential elements. It does not enter into the composition of the food or the composition of the fungus itself. Fungi and actinomycetes require approximately 0.1 to 0.3 ppm of iron. A. *niger* utilizes either Fe^{+2} or Fe^{+3} iron. *Pilobolus* requires iron to be supplied as various chelated iron compounds. Iron is an activator constituent of many enzymes such as peroxidaes, catalases and cytochromes. Other iron-containing metabolites are: (1) coprogen, (2) ferrichromes, the heme pigment of *Neurospora crassa* and *Penicillium notatum,* (3) the pigment, pulcherrimin of *Torulopsis pulcherima,* and (4) spore pigment of black *Aspergillus* spp. which contains 0.20% iron in the pigment aspergillin.

Iron affects the production of penicillin (Koffler 1947). Tandon and Agarwal (1956) recorded that *Gloeosporium psidii* and G. *citrticolum* required 25 *pg/l* concentration of iron to attain maximum growth as well as good sporulation, while a third species, G, *limetticolum,* needed a still higher concentration of iron for best growth, that is, 50 *pg/l,* while sporulation was best only at 5 *pg/l* concentration. *Alternari tenuis* attained best growth at 250 *pg/l* concentration of iron (Grewal, 1956). *Cercosporina ricinella, Colletotrichum gloeosorioides* and *Curvularia penniseti* recorded best growth at 0.2, 0.5 and 0.3 *pg/ml* of iron respectively. Misra and Mahmood (1959) recorded a similar response to iron in *A tenuis* and *Colletotrichum capsici.* Working on *Fusarium* (Saraswathi Devi, 1958) and *Pyricularia oryzae* (Apparao *et al.*. 1955) the essentiality of iron in these two species was demonstrated. The most striking physiological effect of iron deficiency is retarded growth and scanty sporulation. Studies on the trace metal requirements of some soil Fusaria, causing wilt diseases, by Sadasivan and his school (1967) have yielded valuable results. Iron and manganese, particularly the former, were found to be quite effective in altering the potency of fusaric acid by preferring chelating *in vivo* in cotton plants infected with *F. vasinfectum* (Sadasivan, 1965). Iron has been found to influence the pectic enzyme system of the pathogen *F. oxysporum* var. *vasinfectum* (Subramanian, 1956). Iron *in* ferric form produces a striking increase in pectin methyl esterase (PME) activity, even at 0.25 ppm concentration and thus appears to be essential for PME production by this fungus. Iron also increases polygalacturonase activity but only at the lowest concentration. Apparao (I959) showed that *iron* is an effective antagonist against certain toxic ions, such as Cu and Zn, during studies *with pyricularia oryzae.*

Zinc (Zn)

Raulin's studies on *A. niger* (1869) showed that zinc is one of the essential micro-nutrients and since then various investigators have confirmed the necessity of this element (Steinberg, 1919; Roberg, 1928; Foster and Waksman, 1939; Leonian and Lilly, 1939; Blank, 1941; Lindeberg, 1944; Ezekiel, 1945; Yogeswari, 1948; Perlman, 1948 a, and b; Chesters and Robinson, 1951; Tandon and Agarwal, 1956; Grewal, 1956; Saraswathi Devi, 1958; Agarwal and Shinkhede, 1959; Tandon and Chandra, 1962; Sadasivan, 1965, 1967). The adequate concentration in media is 0.5-1.0 ppm. Zinc is an activator component of most of the enzymes, such as carbonic anhydrase. alcohol dehydrogenase, lactic dehydrogenase, glutamic dehydrogenase, etc. Zinc has been shown to be a structural component for ribosomes. High levels of zinc induce genetically stable variants in *Helminthosporium sativum* (Millikan, 1940). Zinc deficiency influences diverse metabolic processes, including organic acid fermentation (Foster and Waksman, 1939; Chesters and Robinson, 1951), aerobic and, anaerobic respiration (Grimm and Allen, 1954; Vallee and Hoch, 1955), aminoacid metabolism (Nason *et al.*, 1951); ascorbic acid synthesis (Kalyanasundrum and Saraswathi Devi, 1955b), nucleic acid metabolism (Nason and Evans, 1953), production of certain antibiotics (Jarvis and Johnson, 1946, 1947; Kalyanasundrum and Swaswathi Devi, 1955a), etc. Zinc concentration in the medium influences the quantity of many primary and secondary metabolites (Bilgrami and Verma, 1978).

Copper (Cu)

Copper is needed in a very small quantity – 0.01-0.1ppm-as compared to other elements. It is very difficult to prepare a medium free of Cu ions. Roberg (I931) made use of Postels method of adding ammonium sulphide to Convert heavy metals to sulphides and absorbing these impurities with charcoal.

Copper is present in many enzymes as tyrosinase, transferring hydrogen from the amino acids. Terminal cytochrome oxidases (cytochrome A and Aa) contain either one or two atoms of Cu per mole of heme.

Steinberg (I936a, 1950) reported that lack of Cu in the media reduced the growth of a number of fungi, including *A. niger, F. Oxysporum, Cercospora nicotianae, Sclerotium rolfsii, Theilaviopsis basicola* and *pythium irregulare.* Agarwal (1959) observed that *F. caeruleum* strains exhibited stimulated sporulation in the presence of Cv, but *Curvularia penniseti* could attain only poor sporulation. Saraswathi Devi (1958) could not establish the indispensibility of Cu for the growth of some soil Fusaria studied by her. Blank and Talley (1941) found that Cu was toxic to *Phymatotrichum omnivorum.* Studies on the pectic enzymes system of *Fusarium vasinfectum* by Subramanian (1956) revealed that at low concentration (10^{-6m}) Cu stimulated both pectin methyl esterase (PME) as well as polygalacturonase (PG) activity. Copper is also known to influence pigmentation on coloured spores of *Aspergillus* spp. (Mulder, 1939) and *Trichoderma viride* (Brian and Hemming, 1950). It may also play a role in the chlorine metabolism of fungi.

Calcium (Ca)

As early as 1922, Young and Bennet observed that *Rhizoctonia solani* could not grow in the absence of Ca and they thus established the essentiality of Ca in the nutrition of fungi. Steinberg (1948) reinvestigated the Ca requirement of fungi and concluded that this element was not required by *Aspergillus niger* and *Fusarium Oxysporum* var. *nicotianae.* However, later work has shown its essentiality for a number of fungi belonging to this group (Cochrane, 1958). Calcium is needed in concentrations of 0.5 ,to 20 ppm. Calcium in many organisms, including fungi, acts as a protection against the injurious effects of certain common monovalent cations, especially H^+, N^{a+} and K^+. It has been found that Ca affects the time required for perithecial formation for *Chaetomium* sp. (Basu, 1951). The intensity of conidiation in *Trichoderma vir ide* is also affected by Ca concentration (Brian and Hemming, 1950).

Manganese (Mn)

Manganese is generally required in small traces (0.005-0.01 ppm). Manganese deficiency has several physiological effects, including a pronounced decrease in sporulation. This metal frequently has a role in many metabolic reactions. Magnesium concentration also affects other enzymes in the cell (Hofman *et al.,* 1950). It has been shown to be an activator of yeast arginase. Manganese is also required in nitrite and hydroxlamine reductase in fungi. Donald *et al.* (1952) observed a 50% reduction in the growth of *A. niger* in Mn-deficient medium. Sulochana and Lakshmanan (1968) recorded a 50 to 60% decrease in mycelium of some strains of *A. niger* in Mn deficient culture medium. Subramanian (1956) observed that the presence of Mn in culture medium in concentration ranging from 0.025 to 0.20 ppm caused the inhibition of pectin methyl esterase (PME) and polygalacturonase (PG) activity, while complete removal of Mn in the medium resulted in a remarkable increase in the PME and PG activity of the mycelium of the pathogen, *F. vasinfectum.* Apparao (1957) and Sadasivan (1967) have shown that Mn can act as an ion antagonist against Cu and zinc toxicity, although with lesser efficiency than Fe.

Molybdenum (Mo)

The essentiality of Mo was first discovered by Steinberg (1936). The metal is needed in an extremely small amount, ranging from 0.1 part per billion to 10 ppb. It is assumed that enzymatic reduction of nitrate is carried out by enzymes which require Mo as an activator. Molybdenum requirements of *A. niger* and *Neurospora crassa* are higher if nitrate and not NH^+_4 is used as the nitrogen source.

Molybdenum is known to be required for the formation and activation of nitrate reductase in fungi. Nicholas and Nason (1954a and b) showed that the nitrate reductase enzymes of *N. crassa* is a flavoprotein which contains Mo.

The role of Mo in the metabolism and growth of fungi growing with nitrogen source other than nitrate, is obscure. It is being realized gradually that this metal has some other role also, possibly in the activity of certain other enzymes. This metal could enhance pectin methyl esterase (PME) and polygalacturonase (PG) in the mycelia of *Fusarium vasinfectum* (Subramanian, 1956).

Other inorganic nutrients, in addition to Zn, Cu, Fe, Ca, Mn, Mo, etc. which are reported to be essential micronutrient are:

Vanadium (V)

A. niger responds slightly to Vanadium (Bertrand, 1941) and fleshy fungi take up Vanadium from the soil.

Gallium (Ga)

Steinberg (1938a) showed the essentiality of this element for *A. niger* based on a purely purification technique of the medium.

Scandium (Sc)

The essentiality of Scandium was again shown by Steinberg (1939a) for *A. niger*, but only when glycerol was the carbon source.

Cobalt (Co)

Although evidence indicating the essentiality of Co has been presented quite often (Marston, 1952) it is doubtful whether it is needed for fungal growth. In bacteria, actinomycetes and animals, Co is known to be essential for the synthesis of compounds of the vitamin B_{12} group (cobalamins); Co forms a constituent part of the tetrapyrrole ring of this vitamin. Fungi have now been shown to synthesize cobalamins (Tanner, 1960). Some fungi are reported to be deficient in cobalamins (Adair and Vishniac, 1958; Littman and Miwatani, 1963).

Phosphorus (P)

This non-metallic element is essential for all forms of life, including fungi. Raulin (1869) recognized the essentiality of P in nutrition of *A. niger*. Phosphorus forms the basic component of nucleic acids (DNA, RNA), the high energy phosphates, ATP, ADP, and the phospholipids. Phosphorus is both a structural and functional element in the growth of fungi.

Fungi apparantly utilize in the form of phosphates. Different phosphates support different amounts of growth. Both inorganic and organic phosphates may be utilized by the fungi.

Fungi exhibit varying requirement of P. Some fungi, such as *Brevilegnia gracilis* (Bhargava, 1945), *Fusarium* spp. (Brown, 1925; Agarwal, 1957; Jaurihar and Mehta, 1972), *A. terreus* (Mehrotra and Tandon, 1970), *Curvularia* spp. (Agarwal, 1958), *Alternaria tenvis* (Singh and Tandon, 1967a, b) etc. are at least able to grow feebly on P deficient media. *Achlya* sp., *Isoachlya anisospora* and *Saprognia* sp. (Bhargava, 1945), and *Fusarium solani, Botryodiplodia ananassae* and *Macrophomina phaseoli* (Bhargava and Tandon, 1963) do not grow at all in the absence of P.

Sulphur (S)

Sulphur is an important structural element in the growth of fungi. The activities of many enzymes depend upon the -SH group. Some vitamins (thiamine, biotin) and nucleoproteins contain S. Volkonsky (1934) designated those fungi that are able to utilize sulphate sulphur as euthiotrophic and those that require reduced sulphur as parathiotrophic. It was observed by many earlier workers that fungi growing on S deficient media either attain feeble growth or fail to survive.

It is now well-known that S plays a significant role in the metabolic activities of fungi. It is also essential for the biosynthesis of sulphur containing aminoacids besides timing a component of SH or thiol groups of many enzymes (Lilly and Barnett, 1951; Cochrane, 1958; Bhargava and Tandon, 1963), co-enzymes and vitamins, which in turn affect various vital processes in fungi, including their choice and ability to utilize various sources of sulphur. Fungi which are unable to utilize 6-valent sulphur (SO_4^{-2}) or which may utilize an organic S source to greater advantage, mostly belong to the aquatic phycomycetes group. Most of the Saprolegniales (Volkonsky, 1933, 1934; Leonian and Lilly, 1939; Bhargava, i94S; Davey and Papavizas, 1962) and all the known members of Blastocladiales (Cantino, 1950), have been reported to be parathiotrophic. *Blastocladia pringsheimii* and *B. ramosa* utilize cystine and cysteine (Crasemann, 1957). They cannot utilize $S0_4$ sulphur.

pythium spp. (Saksena *et al.*, 1953), and *Pestalotia malorum* (Tandon, 1950) are able to utilize a variety of S sources.

Several fungi are also able to utilize sulphur as an element (Abbot, 1923; Sciarini and Nord, 1943; Miller *et al* 1953; Davey and Papavizas, 1962). There are reports that some fungi, such as *A. niger(Steinberg,* 1941), *Curvularia* spp. (Srivastiva, 1951; Singh and Tandon, 1970), *Fusarium solani, Botryodiplodia ananassae* and *Macrophomina phaseoli* (Bhargava and Tandon, 1963) and some Mucorales (Sarbhoy, 1965) can survive without S. Some fungi have been shown to grow without any S source. These are *A. tenuis* (Singh and Tandon, 1967a), *Gloeosporium musarum* (Grewal, 1954), *Pestalotia malorum* (Tandon, 1950), *Fusarium coeruleum* (Agarwal, 1957), *Curvularia penniseli(Agarwat,* 1958), *Phyllosticta* spp. (Tandon and Bilgrami, 1958), *Pestalolia versicolor* (Agarwal and Ganguli, 1960), *Curvuluria pallescens, Alternaria citri* and *A. tenuit* (Hasija, 1969).

CARBON NUTRITION

Carbon compounds have two essential functions in the metabolism of fungi:

1. The synthesis of compounds which make living cells, protein, nucleic, acid, protoplasm cell wall, etc.

2. Carbon compounds are the sole source of energy which account for half the known compounds of all the other elements because of the property of C to form compounds in which one C is linked to another in the form of chains and rings. Hence, a knowledge of C nutrition is fundamental to an understanding of the physiology of fungi. There is much information about the carbohydrates and related compounds as carbon source and the manner in which they are dissimilated and assimilated.

There are various C compounds which are involved in the C metabolism of fungi. These are given in Table 10.1.

The main source of C in the metabolism of fungi comes from carbohydrates. As the very name implies, carbohydrates are a group of organic compounds containing the elements, C, H, O and are generally in the ratio of I : 2 : 1. However, the above criterion for this group has been widened to include compounds containing nitrogen and sulphur and compounds that do not conform to a strict 1 :2: 1 ratio of carbon, hydrogen and oxygen and because of this, carbohydrates are no longer exclusively thought of as hydrates of carbon but are grouped in a more general category as polyhydroxy aldehydes or polyhydroxyketones and their derivates (Devlin, 1975).

Classification: The carbohydrates may be divided roughly into three groups: (1) monosac-charides, (2) oligosaccharides, and (3) polysaccharides.

Monosaccharides and Related Compounds

Monosaccharides have a general formula, $C_nH_{2n}O_n$. The carbon chain is unbranched except in very few cases (sugars). They possess a free (or potentially free) aldehyde (–CHO) or ketone (–CO) group, besides the primary (–CH_2OH) and secondary (–CHOH) alcohol groups. In monosaccharides carbon units, the aldehyde and primary alcohol groups are attached on two extremeties, while the ketone group is located at the secondary carbon atom. In addition, sugars are further

Table 10.1
Classification or More Important Carbohydrates Found in Plants

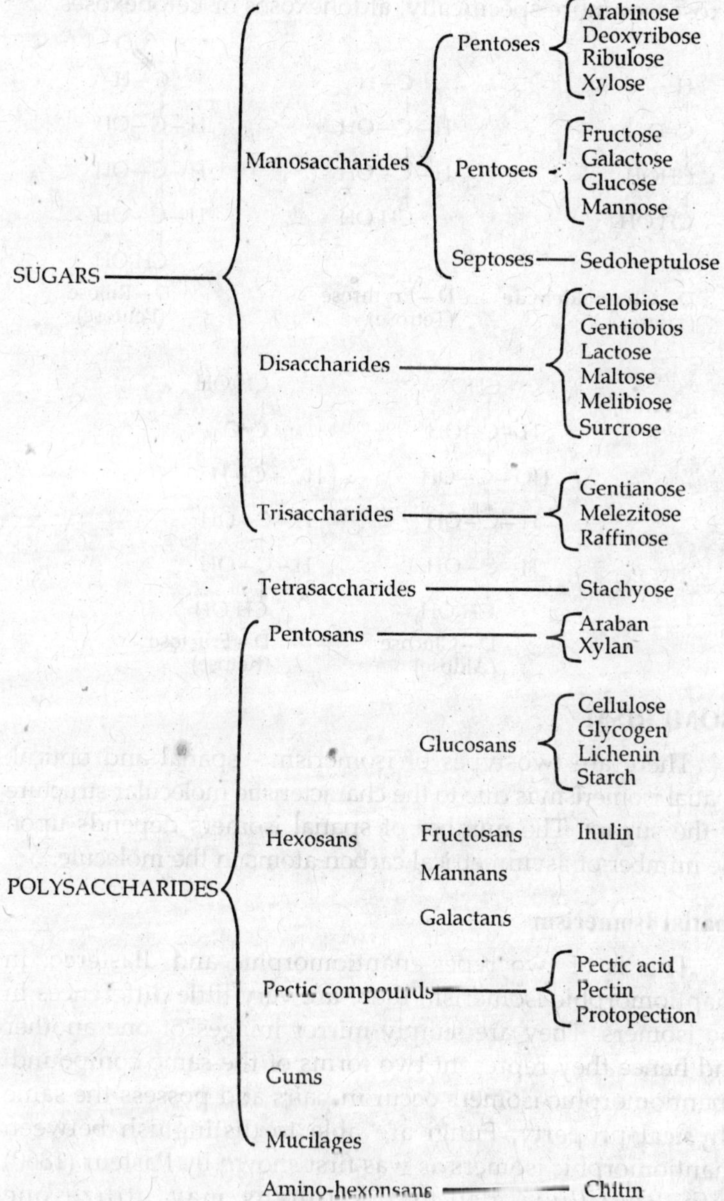

classified according to the number of carbon atoms present and the functional group involved. Thus, there are pentoses, hexoses, or more specifically, aldohexoses or ketohexoses.

```
H               C–H                 O
|               |                  //
C=O         H–C–OH               C–H
|               |                  |
CHOH        H–C–OH           H–C–OH
|               |                  |
CH2OH           CH2OH        H–C–OH
                                   |
                             H–C–OH
                                   |
                                 CH2OH
```

D–Glyceraldehyde (Triose) D–Erythrose (Tetrose) D–Ribose (Pentose)

```
        CHO              CH2OH
        |                |
   H–C–OH                C=O
        |                |
  HO–C–OH           HO–C–H
        |                |
   H–C–OH            H–C–OH
        |                |
   H–C–OH            H–C–OH
        |                |
      CH2OH            CH2OH
```

D–Glucose (Aldose) D–Fructose (Ketose)

ISOMERISM

There are two types of isomerism - spatial and optical. Spatial isomerism is due to the characteristic molecular structure of the sugars. The number of spatial isomers depends upon the number of asymmetrical carbon atoms in the molecule.

Spatial Isomerism

This is of two types-enantiomorphic and diastereo. In enantiomorphic isomerism there are very little differences in the isomers. They are simply mirror images of one another and hence they represent two forms of the same compound. Enantiomorphic isomers occur in pairs and possess the same physical property. Fungi are able to distinguish between enantiomorphic isomers as was first shown by Pasteur (1860) with *Penicillium glaucum*. A fungus may utilize one enantiomorphic while rejecting another. This mirror-image type of isomerism is indicated by the use of the Roman Capitals, D or L, preceding the name of the sugar, that is, D-glucose or L-glucose. A D-molecule is one in which the asymmetric carbon atom farthest from the aldehydic or ketonic group and adjacent to the terminal CH_2O group has its OH group shown on the right. In the L-form, the −OH group of the same carbon atom is shown on the left.

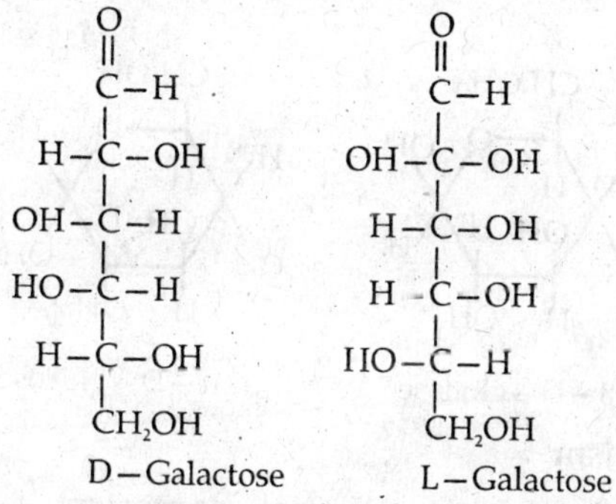

D−Galactose	L−Galactose
O	O
‖	‖
C−H	C−H
H−C−OH	OH−C−OH
OH−C−H	H−C−OH
HO−C−H	H−C−OH
H−C−OH	HO−C−H
CH_2OH	CH_2OH

Fig. D-and L-enantiomorphs of galactose.

In the diastereo isomerism the compounds are not the mirror-images of one another and differ in their physical, chemical and biological properties. They contain the same functional group but in different order. Therefore, diastereo-isomers represent different compounds and are designated

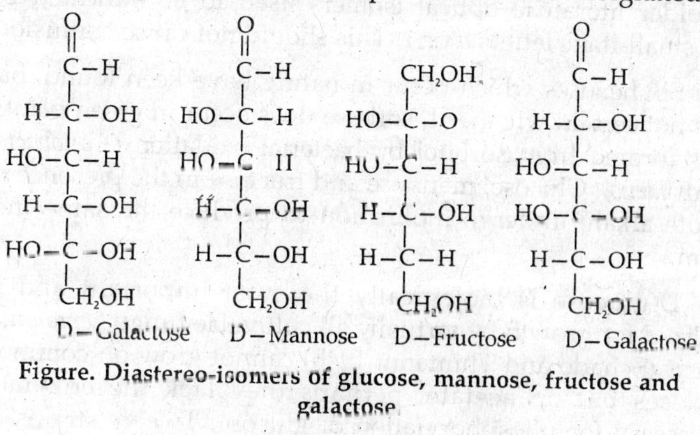

D−Galactose	D−Mannose	D−Fructose	D−Galactose
O	O		O
‖	‖		‖
C−H	C−H	CH_2OH	C−H
H−C−OH	HO−C−H	HO−C−O	H−C−OH
HO−C−H	HO−C−H	HO−C−H	HO−C−H
H−C−OH	H−C−OH	H−C−OH	HO−C−OH
HO−C−OH	H−C−OH	H−C−H	H−C−OH
CH_2OH	CH_2OH	CH_2OH	CH_2OH

Figure. Diastereo-isomers of glucose, mannose, fructose and galactose

by different names, for example, glucose, mannose and galactose are diastereo isomers, as are D-fructose and L-sorbose. The structures are given below.

Two sugars which differ from each other in configuration around a single carbon atom are called epimers. Each of the hexoses exists in *α* and *β* forms depending upon the position of attachment of H and OH groups to the number 1 carbon atom, which is also asymmetric in the pyranose and furanose rings.

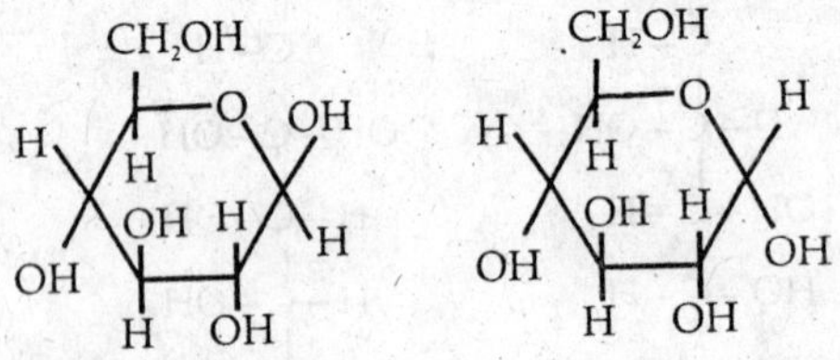

β–D–glucose α–D–glucose

Optical Isomerism

Compounds showing optical isomerism may be either dextro rotatory or levo rotatory, that is, in pure aqueous solution, it may cause a rotation in the plane of polarized light either to the right or to the left respectively. The dextro rotatory are designated by a plus (+) sign and levo by a minus (–) sign in parentheses, prior to the name of the compound, that is, D (+)–glyceraldehyde and D (–)–glyceraldehyde. In older literature optical isomers used to be indicated by the small-italic letters *d* or 1. This should not cause confusion.

Six hexoses which occur in nature have been found, but it is not certain whether L-sorbose does occur in green plants. It is formed from sorbitol by bacterial oxidation *(Acetobacter suboxydens)*. Glucose, mannose and fructose in the presence of dilute alkali, undergo inolination to produce the same inol form.

D-glucose is biologically the most important and is utilized for growth by virtually all cultivable fungi. *Leptomitus acteus* (Schade and Thimann, 1940) cannot grow on common hexoses but on acetate; perhaps they lack the enzymes necessary for phosphorylation of glucose. Twelve strains of

Sacllaromyces have been found which did not of utilize glucose when first isolated but became adapted after slight exposure

```
     CHO               CH2OH              CH2OH
 H-C-OH             HO-C-H                C=O
HO-C-OH             HO-C-H             HO-C-H
 H-C-OH              H-C-OH             H-C-OH
 H-C-OH              H-C-OH             H-C-OH
   CH2OH               CH2OH              CH2OH
D (+) Glucose       D. Mannose         D (−) Fructose

     CHO                CHO               CH2OH
 H-C-OH             HO-C-H                C=O
HO-C-H               H-C-H             HO-C-H
HO-C-H               H-C-OH             H-C-OH
 H-C-OH              H-C-OH            HO-C-H
   CH2OH               CH2OH              CH2OH
D Galactose         L Galactose        L Sorbose
```

Fig.

(Skoog and Lindegren, 1947), Cheo (1949) found *Ustilago striiformis* to grow on glucose only when freshly transferred from a sucrose medium. This is perhaps due to the formation of adaptive enzymes. The hexose sugar sometimes interferes with the utilization of other carbohydrates. This phenome non is referred to as glucose effect. It is believed that the repression of glucose is due to its effect upon certain degradative enzymes. Recently, the phenomenon has been renamed as catabolite repression. Two more mechanisms in the control of carbohydrate utilization by glucose have been reported. These are transient repression (Paigen and Williams, 1966; Moses and Prevest, 1966), and catabolite inhibition (Gaudy *et a.,* 1963; Stumm-Zollinger, 1966) which involves the inhibition of enzyme action and not the formation. This aspect has recently been revised by Paigen and Williams (1970). As indicated earlier, glucose has been reported to be the most efficient source of carbon for most of the fungi which also include the plant pathogens such as *Sclerotium rolfsii* and

Ozonium texanum var. *parasiticum* (Grover and Chona. 1960) and *Pyrhium aphanidermantum* (Kumar and Grover. 1967b).

```
D-Glucose          D-Mannose          D-Fructose
  CHO                CHO                CH2OH
   |                  |                  |
H-C-OH            HO-C-OH               C=O
   |                  |                  |
   R                  R                  R

                    CHOH
                     ||
                    COH
                     |
                     R  common enol form
```

Values of fructose, galactose and mannose have been reported to be almost similar for *Stereum gausapatum* (Herrick, 1940), *Meninoniella echinulata* (Perlman, 1948), *Penicillium digitatum* (Fergus, 1952), *Chalara quercina* (Beckman *et al.*, 1953), *Colletotrichum inamdari* (Hasija, 1965), *Alternaria citri, A. tenuis* and *Curvularia pallescens* (Hasija, 1968, 1970), *Pestalutia sapotae* al).d *P. versicolor* (Agarwal and Agnihotri, 1970). This behaviour has been explained in *Ustilago maydis.* The sexually compatible mono-sporidial lines and their salepathogenic diploid derivatives exhibit varying ability to metabolite L-sorbose. Pentose sugars (D-xylose, D-and L-arabinose, L-rhamnose, and P-ribose) have found only limited application in the nutrition of pathogenic organisms. Their values differ considerably. Some fungi are capable of distinguishing between D and L and enantiomorphs of arabinose. Generally the L-isomers are more. favourable to fungal growth.

Reduction of sugars (-CHO or C=O) yields sugar alcohols, perhaps hexitols, teritols, pentitols and heptitols. Mannitol is more generally utilizable. Sorbitol and dulcitol are unfavourable to most of the fungi. Sorbitol-the reduction product of glucose-is more generaHy used than dulcitol—the reduction product of galactose. Erythritol is rarely used. Mannitol is equivalent to glucose for many fungi and actinomycetes. However, many fungi with generally wide substrate ranges grow poorly or not at all with mannitol as the sole, carbon source. Good to moderate utilization of sorbitol has been reported in *Pythium aphanidermatum* (Kumar and Grover, 1967), and *Phytophthora* spp. (Mehrotra, 1951). However, *Colletotrichum gloeosporioides* (Ghosh and Tandon, 1967), *Helminthosporium rostratum* (Agarwal and Shinkhede, 1959) and *pythium* spp. (Saksena and Mehrotra, 1949) exhibited a poor response to sorbitol. *Phytophthora* spp. (Mehrotra, 1951) can utilize glycerol and erythritol. Tandon and Chandra (1962a) reported poor utilization of glycerol by *Curvularia penniseti*.

NITROGEN NUTRITION

Nitrogen nutrition of fungi has been extensively dealt with by Cochrane (1958), Foster (1949) and Lilly and Barnett (1951). Like carbon sources, nitrogen is also used by fungi, both for functional as well as structural purposes. Nitrogen is one of the most prevalent elements in the living organisms and is found in essential compounds, such as proteins, nucleic acids, some of the plant growth regulators and in many of the vitamins. Fungal cell walls contain chitin which is a polymer of D-acetyl-glucosamine containing nitrogen.

In general, inorganic or organic nitrogen is taken up rapidly during the phase of growth. Robbins (1937) and Steinberg (I939 b, 1950) have classified fungi according to their ability to utilize nitrate, ammonium, organic and elemental nitrogen. The classification is given below:

Table 10.2

Nature of Organism	NITROGEN SOURCE			
	Nitrate	Ammonium Salts	Orange nitrogen	Gaseous Nitrogen
a. Nitrogen fixing organisms	+	+	+	+
b. Nitrate-ammonium utilizing organism	+	+	+	–
c. Ammonium utilizing organisms	–	+	+	–
d. Organic nitrogen utlizing organisms	–	–	+	–

Notes (+) indicates capacity to utilize the source.

(−)indicates incapacity to utilize the source.

Group I includes mostly bacteria, such as *Rhizobium, Azotobacter, Clostridium,* etc. Fungi known for this purpose are very few, for example, *Phoma betae* (Duggar and Davis, 1916) and *P. casuari!la* ('rove *et 01;,* 1949)., There are reports that Plasmo-diophoromycetes (Kaplan and Gotti,1952) is also capable of utilizing atmospheric nitrogen.

Fungi capable of using nitrate, ammonium and organic nitrogen have been placed in group II. This obviously suggests that those fungi which are capable of utilizing nitrate nitrogen can also use ammonium nitrogen. A few fungi have been reported to utilize I nitrate nitrogen, but not ammonium nitrogen. *Cryptococcus nigricans* is reported to be such a fungus (Rich and Stern, 1958).

Fungi placed in group III are capable of utilizing organic nitrogen and ammonium! nitrogen but not nitrate. This class includes most of the fungi which have been studied I when norganic ammonium salts are used as nitrogen sources; the pH of the culture medium I usually falls in many instances to inhibitory levels. This difficulty may be avoided by using ammonium salts of organic acids, for example, diammonium tartarate, or by adding salts of fumaric acid or the other four carbon dicarboxylic acids to a medium containing inorganic ammonium salts (Leonian and Lilly, 1940). The demorcating line between group III and IV is not as sharp as conceived by Robbins.

UTILIZATION OF NITROGEN SOURCES

Fungi Utilizing Nitrate Nitrogen

Nitrates occur in the soil and thus they are a natural source. In general, nitrates have been reported to be an excellent source of nitrogen for imperfect fungi and Ascomycetes (Lilly and Barnett 1951; Thind and Randhawa, 1957; Suryanarayanan, 1958; Agarwal and Ganguli, 1960; Bilgrami, 1964; Narasimhan, 1969; Subramanian and Tyagi, 1968). Inability to utilize nitrate is specially common in some groups, *fot* example higher Basidiomycetes, Saprolegniales, Blastocladiales. as well as two marine Phycomycetes, that is, *Thraustochytrium moiivum* and *T. multirudimentale* (Goldstein, 1963). However, some of these forms may metabolize nitrate nitrogen in later stages if the initial growth is attained at the expense of a readily available nitrogen source. Within a genus, individual species differ in nitrate metabolism. The ability to utilize nitrate nitrogen can be lost by mutation (Raper *et al.*, 1945; Pontecorvo, 1953). The efficiency of different forms of nitrates varies from fungus to fungus. Sodium nitrate, calcium nitrate and magnesium nitrate are inferior to potassium nitrate in the imperfect fungi (Tandon, 1967). Plant pathogenic fungi, like others, also use nitrates very well (Lilly and Barnett, 1951; Thind and Randhawa, 1957). The capacity to use nitrate depends upon the nitrate reductase activity of fungi.

Ammonium nitrate is a1so used extensively by fungi as nitrogen source. Several workers iike Mix (1933), Suryanarayanan (1958), Bilgrami ,1964) and Tandon (1967) have shown ammonium nitrate to be inferior to potassium nitrate for a large number of imperfect fungi. A significant fall in the pH of the medium containing ammonium nitrate during the growth of fungi has been reported by Issac (1949), Pelletier and Keitt (1954) and Srivastava (1955) which is an indirect evidence of the preferential quatilization of ammonium ion. It has been shown in *Scopularfopsis brevicaulis* (Morton and MacMillan, 1954) and *Alternaria tenuis* J (Srivastava, 1955) that nitrate utilization began only when all the ammonium had been exhausted from the medium.

Ammonium salts, such as ammonium sulphate, ammonium carbonate and ammonium cbloride are inferior to ammonium nitrate in general. A rapid fall in the pH of the culture medium has been recorded during the growth of various fungi, such as *Pythium* sp. (Saksena *et af.*, 1953), *Penicillium* sp. (Curtis *et al.*, 1951), *Alternaria tenuis* (Srivastava, 1955) and *Sclerotium rolfsii* (Narasimhan, 1969). The fan in pH due to assimilation of ammonica1 nitrogen results in the cessation of growth of many fungi (Cochrane, 1958; Nicholas, 1963).

A high concentration of ammonia is toxic and toxicity is influenced by many factors such as the nature of the carbon source and the presence of aminoacids, phosphates, etc. Abnormalities in growth, spore germination and development have been traced to ammonium toxicity. Narasimhan (I969) reported that *Sclerotium rolfsii* can grow and utilize ammonical nitrogen even at low pH levels and this. is perhaps the reason for wide geographical distribution of *Sclerotium rolfsii* which is a well-known tolerant of low pH (Subramanian, 1967).

In general nitrite is a poor source of nitrogen for most of the fungi. However, some fungi are able to use nitrite .as the sale source of nitrogen, such as *Phymatotrichum ol}1nivorum* (Blank and Talley, 1941), *Fusarium coeruleum* (Agarwal, 1955), *Botryodiplodia theobromae* and *Diplodia typhina* (Tandon, 1967). Failure to utilize nitrite nitrogen in the acidic pHrange, the m.ost common situation in fungi, is explained on the basis of the existence of nitrites on the acidic pH range in the form of un dissociated nitrous acid which exerts the toxic effect because of its destructive influence on the proteins and amino acids of the fungal cells (Lilly and Barnett, 1951). Several fungi, such as *Fusarium coeruleum* (Tandon and Agarwal, 1953), *Col/etotrichum gloeosporioides* (Thind and Duggal, 1957), 3-anthracnose fungi (Thind and Rawla, 1958), *Cercospora* spp. (Thind and Mandahaf, 1965),. *Drechslera sorokiniana* (Tyagi, 1967), *Curvularia verruciformis* and *Fusarium moniliforme* (Sahni, 1967b), *Cephalothecium roseum* (Thind and Madan, 1969) and *Diplodia natalensis* (Bhatnagar and Pathak, 1970) attained a good growth on nitrite nitrogen if the initial pH was on the alkaline range. Steinberg and Thorn (1942) reported some morphological changes in some fungi on nitrite containing media. On the basis of available data, it could be concluded that in most of tbe fungi, the utilization of ammonium and nitrite compounds is influenced by the pH of the medium.

Organic Nitrogen

Essentially all fungi utilize organic nitrogen although only a comparatively small numbers obtain their nitrogen' in this form. A large number of aminoacids are recorded as good nutritional sources for fungi. Fungi as a group actively decompose proteins in nature to their component aminoacids which can be assimilated. In culture organic nitrogen may be supplied as aminoacids, peptides or peptones. Peptides and peptones yield amino acids on hydrolysis. Fungal species vary greatly in. their response to different amino acids and use some of the latter much more readily than others. '"[his is partly because some fungi meet a part of their carbon requirement also from amino acids~ Various factors, such as permeability, their contamination with vitamins, their enzymatic capacity, etc. are :also attributed as the possible reasons for such varied behaviour. Although the response of different species to any single amino acid varies widely, asparagine most often gives good growth which is generally obtained also with short chain amino acids, such as glycine, Dl-alanine, L-glutamic acid and Dl-aspartic acid. Common amino acid present in the hast generally encourages the growth of pathogenic species of *Phyllosticta* (Bilgrami, 1962). L-cystine, L-cysteine hydrochloride and Dl-me thionine are poor sources of nitrogen in general.

8

Relation of Environment and Nutrition with Disease Development

The weather and soil condition effect the seasonal advancement and geographical distribution of plant disease.

The word 'environment' refers to all the external conditions affecting the life and development of an organism, which include temperature, light and moisture and may also include living factors, such as competing micro-organisms or vectors.

The environment can affect the development of plant disease in different ways. It can affect the perpetuation or overwintering of the pathogen from one growing season to the next, the built-up of both primary and secondary inoculum, the dissemination of inoculum, germination and host penetration. The environment can affect growth and development of the host before being infected in such a way as to affect its susceptibility. Finally, it can influence the actual development of disease after the host has become infected. The development of all parasitic diseases also obviously varies with the pathogen, host, interactions between the pathogen and host and environmental factors in the air and soil.

When we take into account the environment, we have to distinguish between macroclimate and microclimate and between weather in general and the particular weather under which specific diseases develop. For example, temperature and humidity in the general atmosphere may be quite different than in the foliar canopy of dense stands of growing plants.

Influence of Temperature on the Advanced of the Disease

Temperature is undoubtedly one of the most important factors which influences the occurrence and development of many diseases. The late blight of potato is usually considered disease which occurs in the northern latitudes, but it can become destructively epidemic when potatoes are grown during winter in the sub-tropics. Some pathogens develop best in low temperature regions. *Taphrina pruni* and *T. deformans* are confined to the cool, wet parts of the globe. Similarly, *Spongospora subterranea,* the powdery scab, is limited to areas of cold temperatures.

On the other hand, there are diseases which occur in the warm belts of the globe, for example, *Fusarium oxysporum* f. sp. *Lini, F. oxysporum* f. sp. *lycopersici, F. oxysporum* f. sp. *conglutinans* causing wilts of flax, tomato and cabbage yellows.

The seedling blight of maize and wheat are caused by *Gibberella zeae* , but the optimum temperature for the disease is different in the two crops. Blight of wheat is caused at a comparatively higher temperature that is 20 to 28°C. In the case of maize, several disease is caused when the soil temperature is between 16°C and 20°C. It should be mentioned here that wheat normally grows at lower temperatures than maize, and the temperatures which produce the maximum amount of disease of these two crops are actually temperatures (too high for wheat and too low for maize) which are unfavourable for the growth of these two crops. Histological and biochemical studies of roots have shown that at these unfavourable temperatures the cell walls of the roots remain for a considerable time in the primary pectic condition and are easily penetrated by the fungal hyphae. This is a case in which the influence of temperature on the disease acts mainly through the host.

The incidence of cereal rusts depends mainly on humidity and temperature. The incubation period and the germination of uredospores of *Puccinia graminis tritici* is favoured by a temperature of 20°C. In the Indian plains this usually occurs in the period between February and March when the weather just begins to warm up. If the weather becomes cloudy and moist during this period, the rust incidence is immediately high, causing huge damage. According to Stakman and Harrar (1957), the time required for the production of uredospores of *Puccinia graminis tritici* on a susceptible wheat cultivar, can range from 85 days at 0°, to 5 days at 24°C. The incubation period of *Pseudoperonospora humuli* G.W. Wilson can vary from 3 days at 21-25°C to 23 days at 5°C or 11 days at 29°C.

Thlelaviopsis basicola Ferraris which causes black root rot of tobacco is highly damaging when the soil temperature is between 17°C and 23°C but causes little damage when the temperature is between 21 °C and 30°C. This has been attributed to the formation of a phellogen layer at the higher temperature, which arrests the spread of the fungus.

Many virus diseases are more virulent when the temperature is between 20°C and 25°C. The most favourable temperature for disease is between 20°C and 25°C; these include the tobacco mosaic virus, potato mosaic virus, cabbage virus, potato yellow dwarf and others.

Late blight of potatoes does not occur in the plains of India, where potatoes are grown as a winter crop. At this time, the conditions necessary for the growth of fungi do not exist in plains. But in the hills where the crop is grown in the rainy season, late blight is very severe.

Musk melons are relatively resistant to seedling blight and wilt caused by *Fusarium oxysporum* f. sp. *niveum* when soil temperature is at the optimum of 30°C or higher for the growth of melons. However, the disease is abundant and destructive at temperatures below 27°C which is optimum for the growth of the pathogen. The resistance at high temperatures appears to be due to the greater ability of the host to exclude the pathogen from the vascular bundles by the formation of periderm.

Effect of Humidity of Advancement of Disease

Atmospheric humidity and precipitation in the form of rain, fog, dew, etc. determine disease spread of to a great extent. Likewise, the amount of soil moisture also determines the seriously of the disease in soil. The speed of germination of inoculum, its entry into the host and the time of incubation are very much affected by the amount of moisture available. In case of soil diseases the amount of moisture will also affect the soil aeration and the concentration of various salts in the soil.

Downy mildew of grapes is very much done by moist weather and high temperature. Similarly, late blight of potato is favourably affected by high moisture content. A large number of diseases are favoured by increased humidity, such as the late blight of potatoes, Koleroga of arecanuts, downy mildew of cereals, blast of rice, early blight of potatoes, scab of apples and circular leaf spot of coffee *(Xanthomonas malvacearum)*. Precipitation and dew play their own part in increasing atmospheric humidity and cannot be considered in isolation. Heavy dewfall favours the germination of the spores of the pathogen and infections in diseases, such as downy mildew and rusts.

Soil moisture appears to have a main effect on the amount of the big vein disease of lettuce caused by a virus that persists in the soil. Diseases such as the root rot of pear caused by *Aphanomyces euteiches,* the damping off of seedling caused by *Pythium ultimum*, are greatly enhanced by an increase in the moisture content of the soil. But, on the other hand, diseases like common scab of potatoes *(streptomyces scabies)* and white rot of onion *(Sclerotium cepivorum)* usually spread under dry conditions.

Soil moisture appears to have a big effect on the amount of big vein diseases of lettuce, caused by a virus which persists in the soil. The disease is favoured with light moisture content up to 85%.

Karnal bunt of wheat *(Neovossia indica),* which is spread air born, occurs as an epiphytotic only when there is heavy rain and air and soil temperature is around 15-20°C two to three weeks before flowering time. The rain water soaks the spores thoroughly and the low temperature promotes germination. The disease is rare in the absence of both these conditions.

Colocasia blight due to *Phytophthora colocasiae* occurs only in the rainy season especially when there is a continuous spell of moist and cloudy weather. Plants recover as soon as dry conditions return.

The moisture content of the air is usually named as relative humidity. This is defined as the ratio of the actual amount of moisture in the air to the amount of moisture the air can hold if the latter were saturated at that temperature. However, since the amount of water the air can hold varies with temperature (warm air can hold more moisture than cold air), relative humidity is not the best measure of moisture in the air when we are concerned with its effect on the pathogen. A better measure is vapour pressure deficit (VPD). The VPD is measured in terms of pressure of atmosphere and is the difference between the actual vapour pressure of water and the vapour pressure of water in a saturated atmosphere.

Lack of atmospheric humidity is also the reason for the relative absence of certain geographical areas of foliage diseases, particularly those of beans, such as anthracnose *(Colletotrichum lindemuthianum)*, bacterial blight *(Xanthomonas phaseoli)* and *Pseudomonas phaseolicola*. Beans are grown for seed in the arid portions of the USA to avoid seed infections by those organisms which require free moisture on plant parts. Late blight of potato seldom occurs in drier areas, except sometimes in those under irrigation.

The powdery mildew fungi, a group of pathogens, are affected by moisture in an unusual manner. Germination of conidia of the lettuce powdery mildew fungus *(Erysiphe cichoracearum)* greatly decreases at 100% relative humidity (VPD = 0 mm Hg), germinates best at 93% RH (VPD = 2 mm Hg) and germinates to some extent at 0.1% relative humidity (VPD=25 mm Hg). Thus, it is not hard to see why powdery mildews develop in the driest of climates. It is believed that the reason these conidia germinate under such dry conditions is that they contain about 70% water and they carry their own water supply with them.

Soil moisture does not appear to have the same controlling effect on the development of disease as soil temperature. In fact, in judging the effect of soil moisture, it is difficult to consider this factor in isolation from other soil factors such as physical condition, water holding capacity, drainage, retentivity and so on. All these factors tend to make up a microclimate which influences the course of disease collectively and not individually.

Effect of Soil *pH*

The effect of alkalinity and acidity of the soil is so big in many cases as to level the incidence and distribution of certain diseases and to be the most important factor in their control.

Alkaline soils are favourable to the spread of the scab of potatoes. It usually develops between *pH* 5.2 and 8.0 falls sharply when the soil *pH* falls below 5.2. Successful attempts have been made to control this disease by acidifying the soil with sulphur or sulphuric acid.

Other pathogens which grow favourably in alkaline conditions of soil are Texas root rot pathogen, *Phymatotrichum omnivorum,* which attacks more than 1000 spp. of hosts, *Ophiobolus graminis,* the take all of wheat and *Verticillium albo-atrum* vascular parasite of many kinds of plants.

High soil acidity is essential to the development of tobacco root rot caused by *Thielavia basicola,* while such a condition is highly favourable to wilt of tomatoes. The behaviour of *F. oxysporum* f. *vasinfectum* which causes the cotton wilt in the USA and India was studied by Mundkur, who found that the American strain developed better under acidic condition, while the Indian strain developed best in alkaline reaction. Root rot of cotton caused by *Rhizoctonia destruens* is favoured by light soils with acid reactions, while in India Fusarium wilt of cotton develops best in heavy alkaline soils. Covered smut of oats develops at a *pH* value of 7.8 and falls sharply with either a decrease or increase in the soil *pH;* it is moderate at 6.5 *pH.*

Although the *pH* of the soil affects the growth and reproduction of some pathogens directly and this determines disease development, it sometimes is only an indicator of certain other conditions or it acts indirectly by affecting certain biological and chemical processes in the soil. Just as plant pathogens are affected differently by soil *pH,* so many saprophytic soil organisms are also affected. Thus, the *pH* may sometimes influence disease development through its effect on soil microflora. A still more common indirect effect probably is its role in determining the availability of nutrient elements. It is well known, for example, that phosphates, calcium, iron and manganese may be present but in insoluble form and therefore may be unavailable.

Effect of Soil Texture

In the black cotton soils which are clayey and heavy, cotton wilt *(F. oxysporum* f. sp. *vasinfectum)* occurs abundantly and root rot due to *Rhizoctonia bataticola* also occurs. On the other hand, in regions of lighter soils (Gujarat) the case is reversed, that is, only root rot occurs and there is no wilt disease. Both these types of soils are alkaline and the determining influence in this case appears to be the texture (clayey or sandy); *Rhizoctonia* cannot survive poor aeration. It is very sensitive to oxygen tension.

Influence of Light on Advancement of Disease

Although light can influence the development of some diseases, it is rarely a limiting factor in their seasonal or regional occurrence. It probably is not usually a determining factor in the development of epidemics nor in the geographical distribution of diseases.

Light intensity and length of day, however, may affect the survival of inoculum, prepenetration processes, possibly the entrance of some pathogens, length of the incubation period, abundance of sporulation and sometimes the type of pathogenic effect. Light intensity and day length also affect the development of many host plants and can be the determining factors in their viability as media for the growth of pathogens and their relative resistance or tolerance to pathogenic effects.

Although wheat stem rust develops well at light intensities ranging from 500 to 10,000 foot candles, the incubation period is likely to be shorter and sporulation to be more abundant with increasing light intensity. In experiments on *Melampsora lini,* a flax rust, uredia appeared 9 days after inoculation when plants were kept under normal light and after 14 days under reduced light. It has been found that high light is most favourable for leaf rust of rye, *Puccinia rubigovera* var. *secalis.* Also, in experiments by Bever (I934), the incubation period of *Puccinia glumarum* in barley was twice as long at low light intensity as at 960 foot candles intensity and 9 days longer when the barley was exposed to 6 hour days than when it was exposed to 12 hour days.

Light is necessary for photosynthesis and thus the general well-being of the plant is affected by both the quantity and quality of light. Several virus diseases are known to be influenced by light intensity. For example, tomato plants will get infected by sugar beet curly top virus when they are grown under conditions of low light intensity; that is, shaded areas, whereas in fun sunlight they will die. Apparently the concentration of carbohydrate in the leaf affects virus symptom expressions. The intensity of light also affects several fungal diseases. Powdery mildews often develop better under shaded conditions. Although it is difficult to separate the response of the host from that of the fungus to light, it has been shown by Masri and Ellingboe (1966) that low light intensities (25-50 foot candles) favour maturation of appressoria of *Erysiphe graminis,* while high intensities (7200 foot candles) and darkness inhibit appressorial maturation.

The quality of light may also affect plant pathogenic fungi. Leach (I962, 1967) has shown that the ultraviolet portion of the light spectrum initiates and stimulates the sporulation of several plant pathogens, for example, *Botrytis cinerea* and *Helminthosporium avenae.*

Sclerotia of *Sclerotinia sclerotiorum,* when placed in moist saw dust and in the dark, only produce apothecial initials and not apothecia. They develop apothecia as soon as they are exposed to light.

Ultraviolet light and direct sunlight inhibit the germination of the teliospores of *Puccinia graminis.* If the dishes are covered with green light filters, germination normally occurs. *Colletotrichum lagenarium* (anthracnose of cucurbits) produces perfect stages of the culture if exposed to ultraviolet light. Prolonged exposure to ultraviolet light is harmful to fungus for they have fungicidal properties.

Influence of Nutrients in Disease

The relation between nutrition and susceptibility to parasitic and non-parasitic diseases in plants has been recently discussed by Karl Boning (1976). We can distinguish three main types of reactions in the behaviour of the host plants towards the parasite. These are susceptibility, resistance and reduced receptivity. In the case of susceptibility, it is possible for the parasite to penetrate the plant tissue and to spread rapidly without being hindered. The host reacts feebly, or not at all, and under some conditions, the host and parasite may tolerate each other temporarily, while in other cases, the disease follows a systemic pattern which ends in the complete destruction of the leaves attacked by the parasite. In the case of resistance also, the parasite is able to penetrate the plant tissue without any delay in its disease process but its further spread is limited by appropriate counter reactions which mostly consist of necrotization of the tissue attacked. The disease is delayed because its spread is prevented. It comes to a standstill and then remains local and limited to the immediate vicinity of the point of infection. There can, however, also be considerable damage in cases of resistance due to necrotic effects. In the case of reduced receptivity, penetration of the parasite into the tissue is made difficult and is either delayed or does not take place at all. The further spread of the disease is hindered without there being any definite counter reactions in the tissue. The disease is, therefore, again limited to the local point of infection.

These different kinds of reactions to parasite attack are correlated in the following way with the deterrent forms of nutrition. Susceptibility is usually most evident where there is potash deficiency, resistance where there is nitrogen deficiency and reduced receptivity where there is phosphoric acid deficiency. When one nutrient is given in excess there is also a relative deficiency of the two other main nurients. Usually, however, only one deficiency factor predominates and it behaves in line with the excess factor as regards the kind of reaction involved. Thus, one-sided nitrogen excess usually reacts in a susceptible way as in the case of potash deficiency, although, of course, where nitrogen feeding is very one sided, behaviour may also resemble that associated with phosphoric acid deficiency. The relative phosphorus deficiency component acts in a similar way, as resistance and reduced susceptibility frequently have similar effects. Excess phosphorus tends to behave more or less like nitrogen deficiency although the potash deficiency component may also show itself. Generally, we find that nitrogen and potash determine the reaction of the plant to the disease organism more than phosphoric acid does. This may have something to do with the fact that phosphoric acid deficiency represents a combination of two excess factors that act in a contrary manner. This is also true where there is excess phosphorus and where the two deficiency factors, again potash and nitrogen, also have effects.

The effect of different levels of applications of lime is considerably less than that of different levels of nitrogen and potash.

9

Mechanism of Defense in Host

Various plants and animal exists in a common environment but the spread of parasites or action of parasites of one form by another is rare. All plants have resistance to some pathogens under certain conditions. Host-parasite relationships may be inhibited or completely hindered if the host is able to ward off the pathogen by passive or active processes. Disease resistance appears to be the rule rather than the exception in nature. The capacity or ability of a plant to protect itself against a pathogen is governed by its genetic make up and the environmental conditions under which the genes operate. The attributes of the hosts that reduce the chances of infection or the development of the pathogen are considered to be defence mechanisms. Different plants protect themselves against pathogens in different ways. The defense mechanisms of plants against pathogens fall into two categories.

a. Mechanical/Structural defense mechanism.

b. Chemical defense mechanism.

MECHANICAL/STRUCTURAL DEFENCE MECHANISM

The histology of defense in plants has been reviewed by Akai and Dickinson. There is also a recent article by Royle on this subject. The first line of protect against pathogens is the surface boundary which a pathogen must before it can infect the plant. The entry of the pathogens may either be through the epidermal cell walls directly or through the natural openings in the epidermis, such as stomata, lenticels, and hydathodes or through injured areas caused by living or non-living agencies. When once a pathogen has gained entry into the host, its motion and spread is further inhibited between or through succulent parenchymatous cells, and in the vascular diseases through the lumen of the xylem vessels. Certain structural features of the epidermis or its interior may greatly affect the ability of the pathogen to penetrate or to attack a host plant. Such morphological defense structures may be present before penetration or infection or may be produced afterwards as a result of the interaction of the host and the pathogen.

Defense Structures Existing Before Infection

Covering

The cuticle which is formed of cutin and waxes forms the outermost covering of the epidermal cells and appears as a non-cellular, membranous layer. Waxes play a protective role by forming a hydrophobic surface which acts as a water repellent and thereby prevents the retention of water on sloping plant surfaces on which pathogens might be deposited and subsequently further invade the host. The cuticle acts as a physical or chemical obstacle to infection. In *Berberis* spp. infected with *Puccinia graminis,* Melander and Craigie (1927) have attributed disease resistance to the outer epidermal cell wall with a thicker cuticle. There are several examples where cutin acts as a physical barrier and enables a greater degree of resistance. The resistance of strawberry varieties to *Sphaerotheca macularis* was shown by Peries and Jhooty and McKeen to vary with the cuticle thickness. The role of the cuticle in the protection of plants against diseases has been described by Martin .

Epidermal cell Wall

The thick and tough outer wall of epidermal cells forms an important defensive barrier for many pathogens. Lignification or the presence of silicic acid in the epidermis of some plants acts as an important structural protective mechanism. In the case of the rice blast fungus *Pyricularia oryzae.* the outer walls of most of the epidermal cells of the leaves are lignified and the entry of the pathogen inside the host is generally through the thin walled motor cells having pectin rather than lignin. Suherin deposition in potato tuber

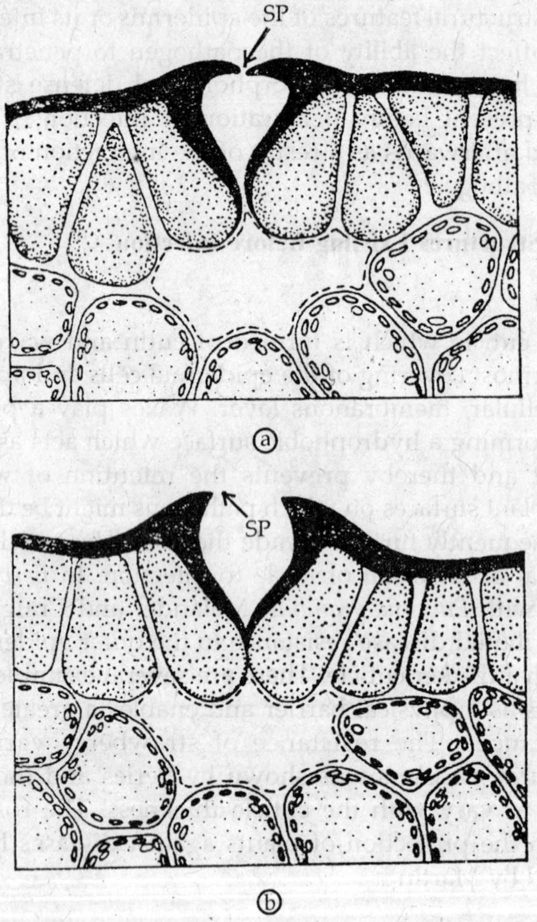

Figure. Stomata of Szinkum mandarin: (a) resistant to the citrus canker bacterium *(Pseudomonas citri)* and of a variety (b) susceptible to the pathogen. SP = stomatal pore.

cells enables better resistance to penetration by some pathogens.

Natural Openings

Most pathogens invades plants through natural openings. In the case of the stem rust of wheat caused by *Puccinia graminis tritici* the Varieties in which stomata open late in the morning are resistant because the germ tubes of the spores germinating in the night dew desiccate. This is due to the evaporation of the dew before the stomata begin to open. This is sometimes called functional resistance. The kind of structure of the stomata of the host gives resistance to penetration in certain pathogenic bacteria. *Xanthomnas citri* is responsible for the citrus canker disease and affects grape fruits and mandarin oranges when artificially introduced into them. The mandarin variety, Szinkum, is resistant to the citrus canker bacterium because it possesses a broad cuticular ridge projecting over the stomata and a narrow slit leading to the stomatal cavity so that the rain drops containing fungal spores and bacteria rarely reach the mesophyll of the leaf (F igure).

Internal Barriers

In certain varieties of wheat, the presence of bundles or increased areas of sclerenchy cells prevent infection in cereal rusts.

DEFENSE STRUCTURES FORMED AFTER INFECTION

Defense structures are formed inside the host plant to prevent the spread of the pathogen such as the formation of cork layers, tyloses and abscission layers; deposition of gums, resins and tannin-like substances; swelling of the cell walls; sheathing of hyphae, etc.

Production of Cork Layers

In the *Rhizoctonia* diseases of potato tubers, following infection, cork layers are formed just below the areas of infection. This layer prevents further attack by the pathogen (Figure). It blocks the spread of toxic substances produced by the pathogen and controls the flow of nutrients and water from the healthy to the diseased area and this results in the starvation of the pathogen. The speed of the formation of the protective barrier is also important. Cork layers are also formed between healthy and diseased areas on *Prunus domestica* leaves attacked by *Coccomyces prunophorae* (Figure).

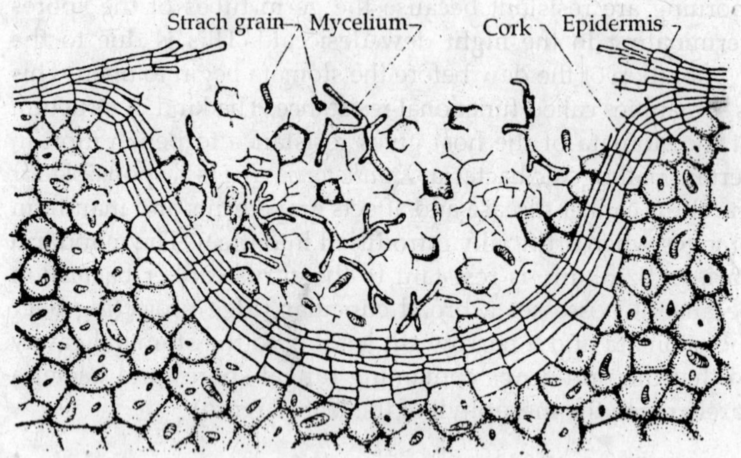

Figure. Formation of cork layer on potato tubers following infection with *Rhizoctonia,*

Tyloses Formation

Tyloses are the overgrowths of the protoplast of adjacent living parenchymatous cells which protrude into xylem vessels through half bordered pits. Tyloses are usually considered to be one of the factors which produce wilt. The time and rapidity of the formation of tyloses determine whether their role will be defensive or whether they themselves will be one of the factors for the causation of the disease (Figure). In sweet potato wilt caused by fungus *Fusarium oxysporum* f. sp. *batatas.* tyloses in some varieties are formed abundantly and quickly before the pathogen and thus being about resistance as this prevents the further spread of the pathogen. Evidence for the importance of vessel tylosis as a defense mechanism is available for vascular wilt of tomatoes.

Formation of Abscission Layers

As a result of infection by several fungi, bacteria or

viruses abscission layers are formed in the young active leaves of stone fruits. Due to the formation of abscission layer (upon infections) in peach leaves by the 'bacterium *Xanthomonas pruni* or the fungus *Clasterosporiurn carpophilum* and in sour

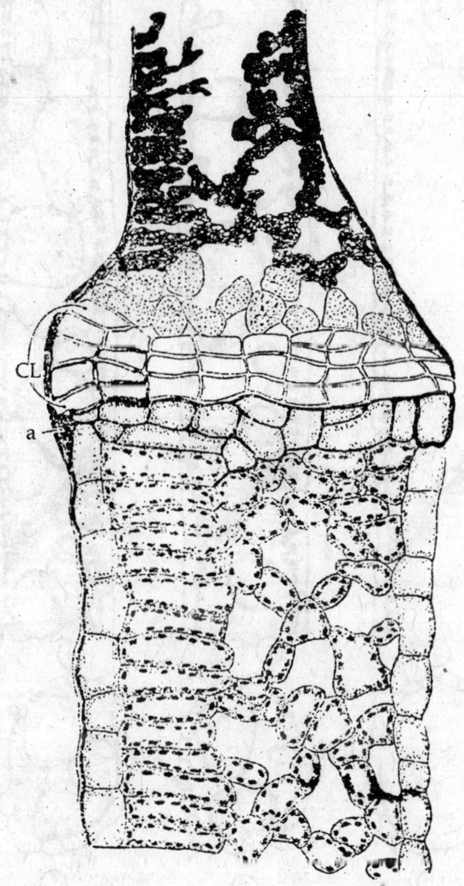

Figure. Formation of cork layer between infected and healthy areas of leaf. CL = cork layer, H=healthy tissue, I = infected, P = phellogen.

cherry trees due to infection with necrotic ring spot and other viruses, there is the swelling of the two layers of cells surrounding the infected spot. They become thin-walled while the pectic materials of the middle lamellae get dissolved which results in the formation of abscission layers (Figure). A shot hole effect results from the development of an abscission layer which extends from the upper to the lower epidermis around

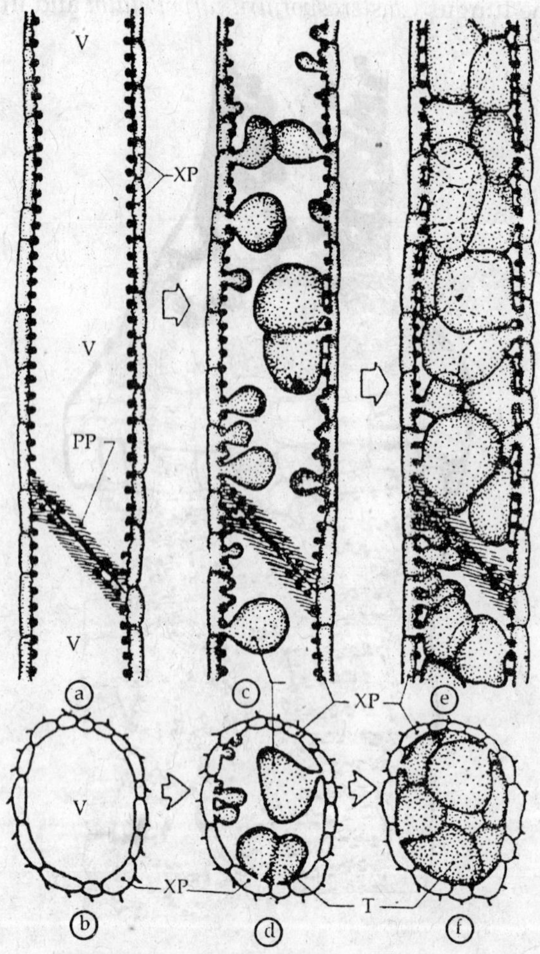

Figure. Development of tyloses in xylem vessels: (a) longitudinal and (b) cross-section view of healthy vessels (c) longitudinal and(d) cross-section of vessels with tyloses (e) longitudinal and (f) Closs-section of vessels completely clogged with tyloses. PP =perforation plate, V = xylem vessel, XP = xylem parenchyma cell, T = tyloses.

the infected tissue and causes the latter to drop out, as in the shot holes of peach caused by *Clasterosporium carpophilum*.

Gum Secretion

Gum deposition along the borders of diseased lesions often serves as a protective demarcation and constitutes a type of mechanical resistance. In the silver leaf disease in plums caused by *Stereum purpureum*, the wood of the host resistant variety, Pershore, produces so much gum in advance of the fungus in a very short time that the fungus becomes completely enclosed by an impenetrable gum barrier in comparison to the susceptible Victoria variety. Hesler (1916) has found a similar gum barrier ill the diseased part of the apple twig infected with *Physalospora cydoniae* (Figure). In varieties of rice resistant to blast disease or *Helminthosporium* leaf spot, gum deposits are formed in the inter cellular spaces

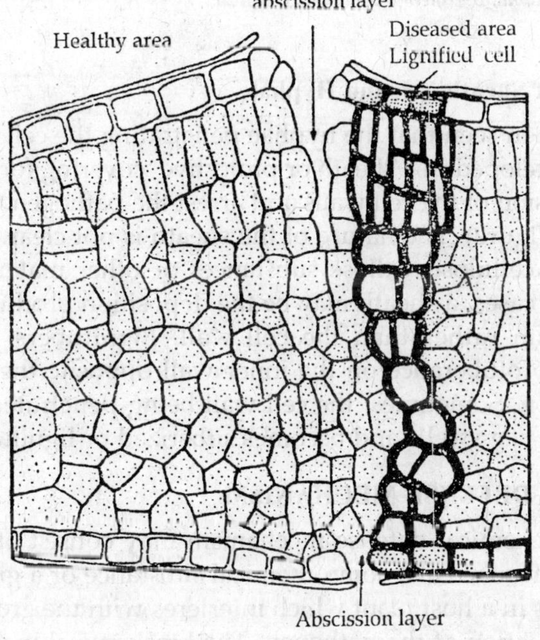

Figure. Abscission layer formation around a diseased spot of a Prunus leaf.

that aid in restricting the fungus to the area of primary invasion These deposits are found to be highly developed in Shoemed rice (resistant to helminthosporiose).

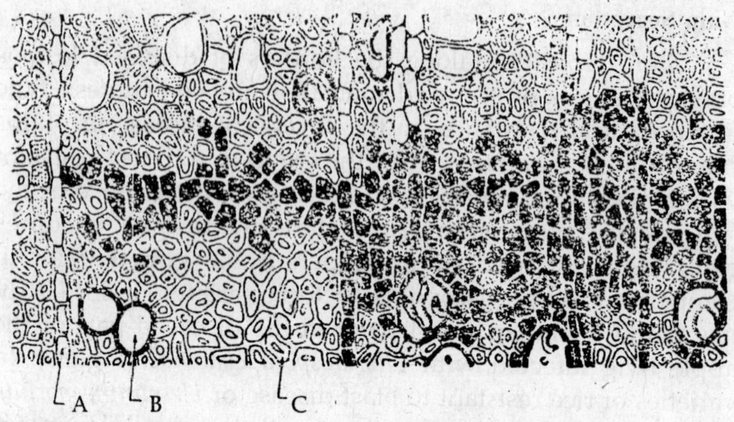

Figure Gum barrier in the apple twig infected by *Physalospora cydoniae*. *V=wood parenchyma, B=xylem vessel, C=wood fibre.*

Formation of Sheath a round Hyphal

It has been seen that the hyphae penetrating the cell wall are often enveloped in a sheath or protective covering formed by the extension of the cell walls and arrest the pathogen from spreading (Figure). The nature of the sheath is not clear. It is made up of cellulose, callose substances or other materials, and in some cases, a sheath may be the deposition product of the cytoplasm rather than the cell wall. Lignitubers may develop as protuberance on the inner wall opposite the site of penetration. Sometimes, the cell wall being penetrated by the infection peg swells and becomes suberized or lignified.

CHEMICAL DEFENSE MECHANISM

The biochemical defense mechanism may consist of the presence or absence of a some chemical substance or a group of substances in a host plant which interferes with the growth and multiplication of the pathogen. The biochemical may be present before infection or may be produced by the interaction of the host and pathogen. This type of defense mechanism which is of a chemical nature is called biochemical defense mechanism.

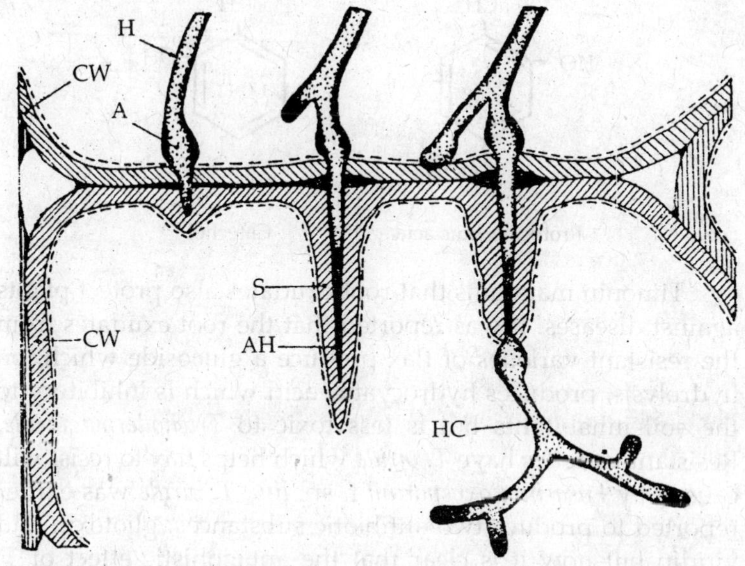

Figure. Formation of sheath around hypha penetrating cell wall. CW = cell wall, H = hyphae, A = appressorium, AH = advancing hypha still enclosed in sheath, HC = hypha in c)'toplasm, S = sheath.

Pre-existing Biochemical Defense Mechanism

Inhibitor Chemicals Released by the Plant in its Environment

Walker and Link worked on onion smudge produced by *Colletotrichum circinans.* Varieties with red scales were resistant in contrast to varieties with white scales. This resistance is due to the fact that the onion varieties with pigmented outer scales contain catechol and protocatechuic acid which are phenolic compounds and which diffuse out from the dead red scales and prevent spore germination. According to Hafiz the blight resistant varieties of *Cicer arietinum* have more glandular hairs which secrete malic acid on the leaves than the susceptible varieties,. A higher concentration of malic acid inhibits spore germination and retards the hyphal growth of the fungus.

Fungitoxic substances have been detected in plant waxes, and cutin acids of citrus lime are toxic to *Gloeosporium limetticola.*

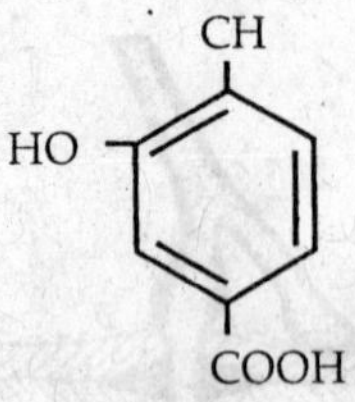

CH HO COOH

Protocatechuic acid

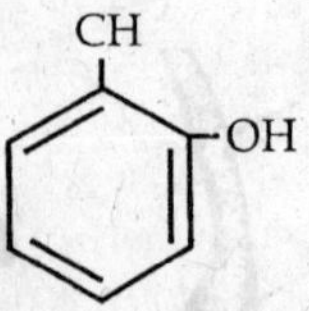

CH OH

Catechol

Timonin maintains that root exudates also protect plants against diseases. It was reported that the root exudates from the resistant varieties of flax produce a glucoside which, on hydrolysis, produces hydrocyanic acid which is inhibitory to the soil inhabitants but is less toxic to *Trichoderma viride.* Resistant varieties have *T. viride* which helps flax to resist wilt caused by *Fusarium oxysporum* f. sp. *lini. T. viride* was earlier reported to produce two antibiotic substances, gliotoxin and viridin but now it is clear that the antagonistic effect of *T. viride* is not due to these two antibiotic substances which, in fact, it does not produce. The antibiotic action of *T' viride* is due to the synthesis of volatile antibiotic substance known as trichodermin.

Inhibitory Substances Present in the Plant Cells before Infection

In potato scab caused by *Streptomyces scabies,* higher concentrations of chlorogenic acid have been reported in potato tubers and in citrus lime leaves.

Turner reported that oat leaves and roots contain a fluorescent glucoside (avenacin) which is end to stop to the growth of several fungi. The growth of *O. graminis,* the cause of take all disease of wheat, is also checked by the sap exuded from the oat roots, but growth of *O. graminis* var. *avenae* is not. Yet the purified inhibitor is equally both the poisonous strains. On the other hand, Turner has shown that a specific, glucoside (avenacinase), produced by the strain var. *avenae,* pathogenic to oats while the wheat race completely inhibited by 5 μ g/ml avenacin, lacks this enzyme. This enzyme destroys the biological activity of the inhibitor and hence the fungus is able to infect the host. Thus, Pathogenic ability and specificity of O. *graminis* var. *avenea* to oat is dependent on the inactivation of the inhibitor.

In case of resistance of pears to fire blight caused by *Erwinia amylovora,* that resistance in certain tissues is related to the presence of phenolic glucoside arbutin. This is hydrolyzed by β-glucosidase of host origin to yield glucose and an antimicrobially active semiquinone. Arbutin is evenly distributed in the tissues of the pear tree, yet these tissues vary in their susceptibility to *E. amylovora.* investigators positively correlated those portions of the tree that were resistant (floral bracket) with the high levels of β-glucosidase activity. Likewise, those parts of the tree that were most susceptible (nectaries of flowers) had decreased levels of β-glucosidase activity.

According to Kuc, phloridzin, the glucoside of phloretin, is present in the leaves of both resistant and susceptible apple varieties. There is no correlation between the compounds of phloridzin and resistance to pathogens.

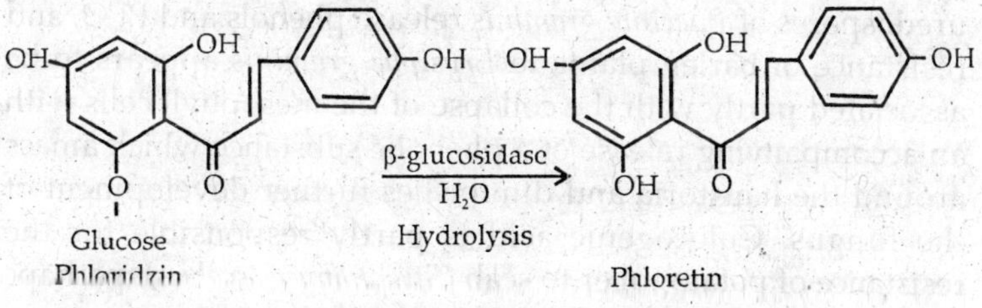

OH OH OH OH OH OH
O O OH O
Glucose
Phloridzin
β-glucosidase
H_2O
Hydrolysis
Phloretin

The hydrolysis of phloridzin to phloretin by β-glucosidase.

When the pathogen invades the leaves of a resistant variety, the plant cells around the invasion collapse and die. Apparently, the non-toxic glucoside phloridzin is hydrolyzed by a β-glucosidase to yield phloretin and glucose. The phloretin in turn is oxidized by phenoloxidases to yield highly fungitoxic compounds. In susceptible varieties, the fungus penetrates the leaves and grows extensively below the cuticle for 10 to 14 days without starting the collapse of the host cells. Thus, phloridzin is not hydrolyzed and the pathogen is not contained.

Production of Phenolic Substances

Resistance to several fungal plant pathogens has been ascribed to higher concentrations of fungitoxic phenolic substances and their oxidation products and to enhanced polyphenoloxidase (PPO) activity which generally, but not invariably, results from infection.

The activity of polyphenoloxidase (PPO) seems to be vital because it can oxidize phenolics to quinones which may be more fungitoxic. Polymerization into dark-coloured melanins may then occur. PPO produced by the pathogen may oxidize the host polyphenols to more highly fungitoxic substances which may avoid further development of the pathogens. For example, a PPO inhibitor decreases the toxicity of catechol to *Cochliobolus miyabeanus* (leaf blight of rice) and prevents melanin development, presumably by preventing enzymic conversions of catechol into fungitoxic quinones. The melanin is apparently fungitoxic although in other cases these insoluble pigments are relatively non, toxic to fungi. Germinated uredospores of *Puccinia graminis* release phenols and PPO, and resistance of barley plants to *Erysiphe graminis* appears to be associated partly with the collapse of the mesophyll cells with an accompanying release of a phenolic substance which amass around the haustoria and diminishes further development of the fungus. Chlorogenic acid is partly responsible for the resistance of potato tuber to scab *(Streptomyces scabies),* perhaps by stimulating protective cambium activity.

Hare points out that aromatic substances, such as polyphenols, phenolic glucosides, flavonoids, anthocyanins, aromatic amino acids tend to amass in and around infected plant' tissue and also in tissues adjacent to wounds where presumably they might exert a fungistatic effect. Phenolic compounds 'are produced in plants primarily via the shikimic and acetic acid pathways. In the latter, phenolics are produced by a head to tail condensation of acetate units derived from the breakdown of sugars during respiration. In the shikimic acid pathway, phosphoenol pyruvate from glycolysis reacts with erythrose produced during the pentose pathway, the activity of which is increased in diseased plants, and forms dehydroquinic and then shikimic acid. From the shikimic acid and various intermediate compounds, some of which are not, phenolic compounds are produced. Some of the most important phenolic compounds involved in the defense of plants are caffeic acid, ferulic acid, chlorogenic acid, phloretin and various phytoalexins.

The high concentrations of chlorogenic acid in the roots of certain potato varieties has been considered the main mechanism of defense against the *Verticillium* wilt pathogen. Resistant varieties contain more chlorogenic acid than the susceptible younger roots and are thus not attacked, but they become susceptible later on. Chlorogenic acid is a competitive inhibitor of IAA oxidase and may be involved in diseases where resistance is increased with high IAA content of the plant. Chakravarty and Srivastava have attributed the resistance of carrot roots to *Pythium aphaniifermatum* to an unidentified, apparently preformed phenolic substance.

Allen and Kuc presented testimony to the fact that sterols may be involved in defense mechanisms. The steroid glycoalkaloids, *a* -solanin and *a* -chaconine, were the major fungitoxic. substances in potato peel, being present in concentration by at least eight times the amount necessary to inhibit 50% of the growth of *Helminthosporium carbonum,* a fungus which does not attack potato. *Phytophthora infestans* is not affected due to the lack of cell-membrane sterols, one of the two prerequisite for susceptibility to saponins. Tomatine, another steroid glycoalkaloids of solanaceous plants, was more toxic to non-pathogens than to pathogens of tomato (Allen and Kuc' 1963) and *Septoria lycopersici* (tomato leaf spot) was able to detoxify tomatine, whereas other *Septoria* spp., non-pathogenic to tomato, were unable to do so. These results provide some circumstantial proof that tomatine may protect tomato leaves against certain pathogens and further investigation is needed. The mode of action of these fungitoxic steroidglycoalkaloids is not properly understood.

Starvation of the Pathogen

A pathogen will only grow and infect a host plant only if only nutrients are provided by host tissues. Walker and Stahmann suggested that in, such pathogens the availability of highly elaborate growth factors, such as vitamins, polypeptides, amino acids and proteins, and the presence of complex enzyme systems may be important for successful parasitism, and that resistant plants may lack one or more of these factors. Some mutants of *Venturia inaequalis* which were deficient in choline and riboflavin were not able to cause scab unless these substances were supplied. A total of 54 biochemical mutants including those in which production of various vitamins, organic bases and amino acids has been impaired have been studied in *Venturia inaequalis* and the effect of each mutation on the pathogenicity of the fungus has been determined of the vitamins requiring mutants only those in which metabolic reactions involving choline or riboflavin have been affected, were found to have lost in pathogenicity. All mutants in which the metabolism of purines, pyrimidines and amino acids had been affected, incited only the fleck reaction and were non-pathogenic.

Resistance has also been attributed to the concentration of nutrients in plant tissues, as different from the absence of specific nutrients needed by the pathogen. Sugar content is perhaps the best known example of this. High sugar content of the tissues favours some pathogen, such as certain rusts and powdery mildews, whereas the reverse is true for others, including *Alternaria solani* (early blight of potato) and *Ceratostomella ulmi* (Dutch elm disease), as discussed by Horsfall and Diamond. Plants become resistant to high sugar pathogens when the sugar content of their tissues falls to a certain level and this may be brought about for example, by treatment with certain growth regulating substances. Boron facilitates the translocation of sugars so that these tend to amass in the leaves of boron deficient plants and the latter are apparently resistant to *Cochliobolus sativus* (barley leaf spot), a low sugar pathogen, but susceptible to *Erysiphe graminis*, a high sugar pathogen.

Beever found that the xylem sap of trees (peach, plum) susceptible to *Stereum purpureum* had more nitrogen and carbohydrate and was more favourable for the growth of the fungus than was the sap of resistant trees (apple, pear). There were also qualitative differences in that the main sugar and sugar alcohol in the sap from resistant trees were sucrose and sorbitol, neither of which supported good growth of *S. purpureum,* whereas the sap of susceptible trees also contained glucose and fructose which were more favourable for growth of the fungus.

Osmotic Pressure and Permeability Effects

According to Lilly and Barnett, the cells of parasitic fungi have more than usual unusually higher osmotic pressure than that of the surrounding host cells, this being necessary for absorption of water from the latter. In the case of *Erysiphe cichoracearum* , cells of resistant tissues or varieties have a higher osmotic pressure than those of the susceptible ones. Further, resistance can be modified by changing the mineral nutrition of the plant and shaded leaves or those cultured on distilled water become susceptible, whereas young lettuce seedlings with high osmotic pressure are resistant (Schnathorst, 1959).

Defense through Absence of Common Antigens

The gene-far-gene theory of the relationship between the host and pathogen evolved from the work of Flor with flax rust. The idea that the antigens shared by the host and pathogen may determine the nature of disease reactions has been explored by Doubly. Tests of globulin antigens from four faces of rust and four from near isogenic lines of flax gave the idea that susceptibility depended on the presence, in the host, of a protein serologically similar to the one possessed by the pathogen. Pathogenic races of *Xanthomonas malvacearum* which cause the angular leaf spot of cotton, had more antigens in common with those of cotton leaves than did the non-pathogenic races of bacteria and the black rot fungus *Ceratocystis fimbriata* which attacks sweet potato.

Phytoalexins

Long lasting immunity against certain virus diseases in humans and animals has been known to science for many years. However, there has been no evidence that plants possess anything like an antibody-forming mechanism which makes for recovery and protects against subsequent infections as in animals. The virus diseased plants do not generally recover and in fact they remain infected till they die. Wingard for the first time, presented evidence that there were symptoms of recovery in tobacco ring-spot-virus-infected plants. Wingard's view that these infected plants acquired an immunity to the disease was vigorously supported by Price. Later, scientists opposed the use of the term immunity to this phenomenon of recovery, usually because the disease clearly had two phases -the acute and chronic. Moreover, the recovered plants contained the virus. Indeed, Wallace suggested that in the case of tobacco plants infected with the sugar beet curly top virus, recovery was due to the *in vivo* development of protective substances and considered that they could be transferred to other plants where they would confer a form of passive acquired immunity. This suggestion put forward by Wallace was followed soon after by a somewhat similar approach to a host reaction in the potato- *Phytophthora* disease resulting in the discovery of substances called phytoalexins (Muller and Boerger, 1940). During the last 30 years many papers have described the structure synthesis and activity of such compounds, now called phytoalexins. The term was introduced by Muller and Boerger (1940) to describe the substances that inhibit fungus development formed when living plants tissues are invaded by a fungal parasite. It has since been redefined and broadened to include substances formed by host tissue ". . . . in response to injury, physiological stimuli, infectious agents or their products (that) accumulate.. . . . to levels which inhibit the growth of micro-organisms" while Day (1973) proposed to include substances that are formed in response to infection that are toxic to, or which repel insect or nematode parasites. Phytoalexins can be described as antibiotics which arise from metabolic interactions of the host and parasite and which are inhibitory to micro-organisms, insects or nematodes which attack plants. They may be attached with host cell necrosis (hypersensitivity) at the infection site out not necessarily so when induced by chemicals. It was envisaged that the factor which induces the formation is produced by the parasite into the host cell. Phytoalexins are now being increasingly and have been detected in several diseases, but more investigation of their overall significance as a resistance mechanism is needed. There are two hypotheses put forward to explain the mechanism of the induction of phytoalexins.

A. De novo synthesis of enzymes, or

B. activation or activity control of enzymes already formed.

De Novo Production of Enzymes

Hadwiger and Schwochau proposed a gene activation hypothesis for pisatin biosynthesis in *Pisum sativum on* the Jacob and Monod model for gene activation in bacteria. Operon activity was presumed to be checked by a regulator gene which continuously synthesizes a specific short-lived repressor. It was visualized that in uninfected and untreated tissues, the operator site is occupied by the repressor, thereby inhibiting the expression of the structural genes, whereas in infected or chemically treated tissues, the structural genes are freed owing to the impairment of either the function or the formation of repressor compounds. The induced pathway of phytoalexin production was postulated to commence at the conversion of phenylalanine to trans-cinnamic acid catalyzed by phenylalanine ammonialyase.

Activation Control by Pre-formed Enzymes

A. Control by naturally occurring protein or peptide inhibitors, a number of peptide-enzyme complexes have been isolated from animal tissue.

B. Another kind of controlled mechanism that could be envisaged is through the structure of the enzyme molecule itself which is known as the allosteric mechanism.

In addition to fungal metabolites, phytoalexins may be induced by heavy metal ions. metabolic inhibitors, some amino acids, a wide range of antagonists of nucleic acid and protein synthesis.

The conclusions of Muller and Boerger (1940) are given below:

a. A principle designated as phytoalexin which inhibits the development of the fungus in hypersensitive tissues is formed or activated only when the host cells come into contact with the parasite.

b. The defense reaction occurs only in the living cells.

c. The inhibitory material is a chemical substance and may be regarded as a product of necrobiosis of the host cell.

d. The phytoalexin is non-specific in its toxicity towards fungi; however, fungal species may be differentially sensitive to it.

e. The basic response that occurs in resistant and susceptible hosts is similar. The basis of differentiation between resistant and susceptible hosts is the speed of formation of phytoalexin.

f. The defense reaction is confined (the tissue colonized by the fungus and its immediate neighbourhood.

f. The resistant state is not inherited.

Phytoalexins are probably fungistatic rather than fungicidal but this depends partly on the quantity present.

According to Cruickshank, the rate of formation of phytoalexin is probably crucially important in combinations of host variety parasite strain (as in the reaction of the different varieties of potato to different races of *Phytophthora infestans)* but probably less so when combinations of pathogens and non-pathogens of specific hosts are involved. Host specificity of phytoalexins occurs between genera but probably not within a species or even between related species. Gene-for-gene relationship might be interpreted by the phytoalexin characteristic of the pathogen, both being genetically determined; a range of host-parasite interaction types might thus arise. Susceptibility could be related to the failure of the invading pathogen to stimulate the formation of phytoalexins or to its tolerance of the amounts synthesized, which in turn is related to its ability to degrade the phytoalexin.

There is increasing evidence that some of the fungi which attack a plant are able to do so because they have the capacity to degrade its phytoalexin and the non-pathogens are not able to do this. Relatively few phytoalexins have been investigated for their contribution to varietal resistance. From the point of view of practical agriculture, the mechanism of varietal resistance and susceptibility are much more pertinent than a fundamental consideration of immunity.

Ipomeamarone ($C_{15}H_{22}O_3$ mol. wt. 230)

This is an abnormal sesquiterpinoid (tetrahydrofuran) which seems to be induced in sweet potato *(Ipomea batatas)* tissue infected with black rot fungus *Ceratocystis jimbriata* by a possible disturbance of carbohydrate and fatty acid metabolism of the host. These authors studied that the biosynthesis of ipomeamarone may involve mevalonic acid as a precursor and could perhaps also be induced by poisonous chemicals and pectinase. The presumption was that the carbon sources being utilized for respiration and biosynthesis of phenolic compounds, may, by an inhibition of the TC-A cycle and pathway of the biosynthesis of phenolics, be diverted to the biosynthesis of ipomeamarone. The mechanism of respiratory increase in connection with polyphenol formation in response to infection, has been studied in detail. ATP utilizing systems are increased and increasing polyphenolic compounds are induced by an increased supply and accumulation of the precursor molecules of C_a, C_4, C_5 and C_6. Ipomeamarone increases substantially in diseased tissues adjacent to infected tissues. Such tissues show terpenes which are also produced by slices of sweet potato in response to a chemical stimuli. The interesting part is that ipomeamarone is toxic to both non-pathogenic and pathogenic fungi and to some bacteria, and there seems to be no correlation between its inhibitory effect and pathogenicity.

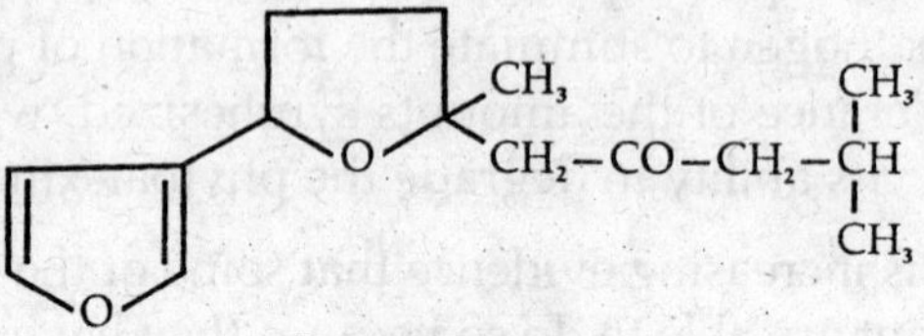

Ipomeamarone

According to Uritani, the carbon sources used in the infected sweet potato for the respiration and biosynthesis of phenolics may be diverted to the synthesis of ipomeamarone by the inhibition of the TCA cycle and phenol biosynthesis. Ipomea-marone was identified as 2-methyl- 2 (4-methyl- 2-oxypentyl) - 5 - (3-furyl)-tetrahydrofuran.

Pisatin

Pisatin has the chromocoumarir ring system and is a phenolic ether ($C_{17}H_{14}O_6$ mol. wt.314).

This is synthesized by the exposed endocarp of detached pea pods in response to inoculation with many fungi or injury. When different fungi, pathogenic and non-pathogenic to pea, are used for inoculation, they induce the production of different concentrations of pisatin. The non-pathogenic fungi induce the formation of pisatin at concentration high enough to stops their growth. If pea pathogens induce the formation of pisatin, it is at concentrations much below these that are toxic to the pathogen. Production of pisatin by pea pods inoculated with *Monilinia fructicola*, a non-pathogen, is reduced or inhabited at high temperature on anaerobic storage, and the pea pods become susceptible to this fungus, while storage at normal temperature conditions results in the accumulation of high levels of pisatin and resistance.

Pisatin can also be formed when dilute solutions of certain metallic salts (especially those of mercury, silver and copper) are put on pea pod endocarp tissues and it is possible, that phytoalexins are produced when metallic fungicides are applied to crops. Certain metabolic inhibitors, such as sodium fluoride and sodium monoiodocetate, are reported to have same effects. Other substances reported to induce pisatin production by pea pods include ethylene which at a concentration of 1 ppm induces a slight production. At certain concentrations, a number of antibiotics (actinomycin D, cycloheximide, chloramphenicol and others) and ribonuclease stimulate pisatin formation; however, they not so at higher concentrations. Many of these substances interfere with protein synthesis in plants.

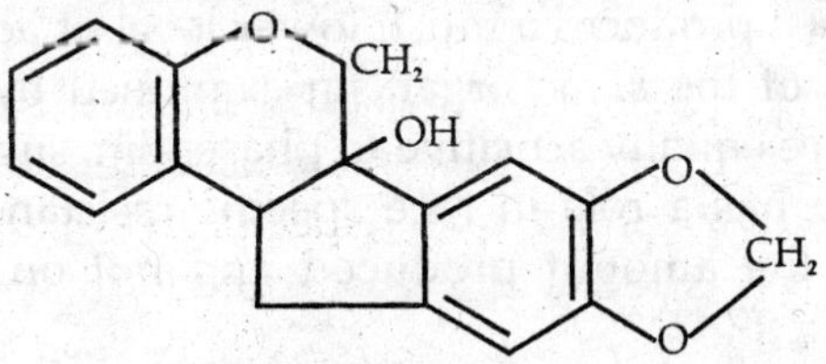

pisatin

As reported by Ruscoe, the production of pisatin diminishes as pea leaves become senescent. He suggested that the increased chance of infection of such leaves to the same fungi stems from the inability of senescing leaves to produce an effective amount of phytoalexins. The knowledge that treatment of leaf discs with benzyladenine reduces senescence and also maintains pisatin production, would give weight to this conclusion. It may be that the increased resistance of leaves to some pathogens which follows the application of benzimidazole and kinetin—substances which maintain protein synthesis and so delay senescence -is due to a continued synthesis of phytoalexin. Such treatments might have potentialities for delaying senescence in plants and thus reducing their susceptibility to infection by pathogens, such as *Botrytis* which often attack senescent tissues. There is increasing proof that some of the fungi which attack plants are able to do so because they can degrade its phytoalexin, and that non-pathogens are unable to do this.

Phaseolin

Phaseolin is almost same to pisatin in chemistry and function. It was separated from detached, open bean pods, following inoculation into the non-pathogen *Monilinia fructicola*. More recently, a number of compounds structurally similar to phaseollin have been identified from diffusates or tissues inoculated with fungi, bacteria and viruses. These are phaseollidin and phaseollinisoflavan and kievitone. According to Rahe phaseollin is important in determining varietal resistance to *Colletotrichum lindemuthianum*, the cause of anthracnose of beans. It accumulates most rapidly in bean hypocotyls inoculated with an avirulent race. Inoculation with a virulent race produces a much lower level of accumulation. Three types of the same organism examined by Bailey and Deverall were equally sensitive to phaseollin, suggesting that if phaseollin has a role in race specific resistance, it would depend on the amount produced and not on differential toxicity.

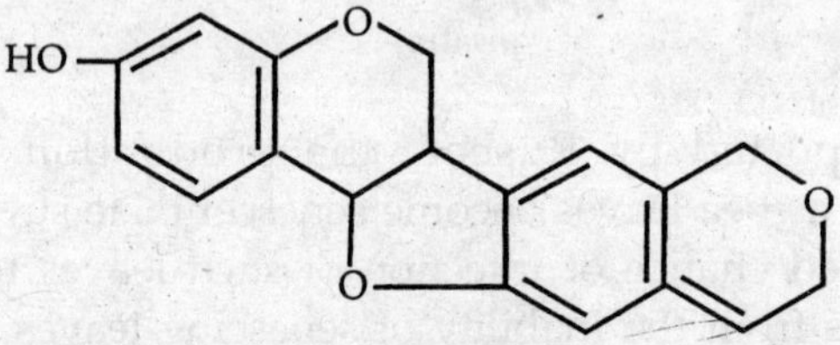

HO O O O

Phaseollin

Phytophthora megasperma var. *sojae* is a root parasite of soybeans *(Glycine max)* Which can be controlled by oligogenic race specific resistance. Keen set out to confirm earlier reports that resistance resulted from the production of a phytoalexin. They showed that 6-a hydroxyphaseollin (HP) was formed in hypocotyls of seedlings of the resistant variety, Harosoy 63, within two or three days after inoculation with mycelium of an avirulent race of *P. megasperma sojae*. The near-isogenic susceptible variety, Harosoy, inoculated with the same race, also produced HP, but at rates that were at least 10 times lower.

Both pisatin and phaseollin have the chromocumarin ring system. Pisatin is produced jointly by the acetate-malonate and shikimic acid pathways.

Trifolirhizin ($C_{22}H_{22}O_{10}$, mol. wt. 446)

This is a new glucoside which has been isolated from the roots of red clover. Its aglycone has the formula, $C_{16}H_{12}O_5$' Trifolirhizin is of interest here, as its structure indicates that it is chemically closely related to pisatin.

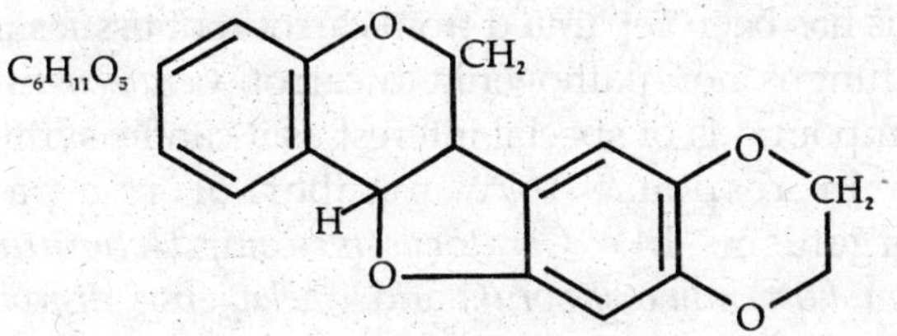

$C_6H_{11}O_5$ O CH_2 H O O CH_2 O

Trifolirhizin

Orchinol ($C_{16}H_{16}O_3$, mol. wt. 256) and Hircinol ($C_{15}H_{14}O_3$, mol. wt. 242)

Orchinol is a phenanthrene and one of the three compounds formed by the interaction of diffusible metabolities of *Rhizoctonia repens* with *Orchis militaris* tuber tissues. According to Gaumann, when there is a stimulus from outside, both *Orchis militaris* L. and *Loroglossum hircinum* (L.) Rich synthesize at the same time two substances orchinol and hircinol. *Orchis militaris* produces 50 to 100 times more orchinol than hircinol, whereas L. *hircinum* produces 100 times more hircinol than orchinol. Both substances are present in tissues of bulbs, *roots* and stalks under aerobic conditions, only as a defense mechanism from specific fungal infection, especially mycorrhizae, and not against common saprophytic or semiparasitic soil fungi. Gaumann considers that the

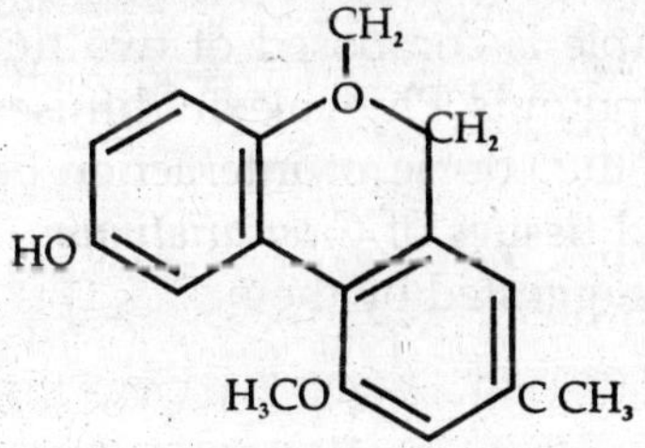

CH_2 O CH_2 HO H_3CO C CH_3

Orchinol

stimulus causing the synthesis of hircinol is a microbially induced antigenic effect resulting in antibody production or from lesions, a traumatic induction of antibody production. Antimicrobial spectrum of orchinol is broad and hardly specific.

Isocoumarin (CUH1204, mol. wt. 208)

This has been separated from carrot root tissues inoculated with a fungus non-pathogenic to carrot, *Ceratocystis fimbriata.* This compound is of special interest as it can be synthesized in tissues in response to a number of non-pathogenic microorganisms like *Ceratocystis ulmi, Helminthosporium carbonum, Fusarium oxysporum* and *Thielaviopsis basicola.*

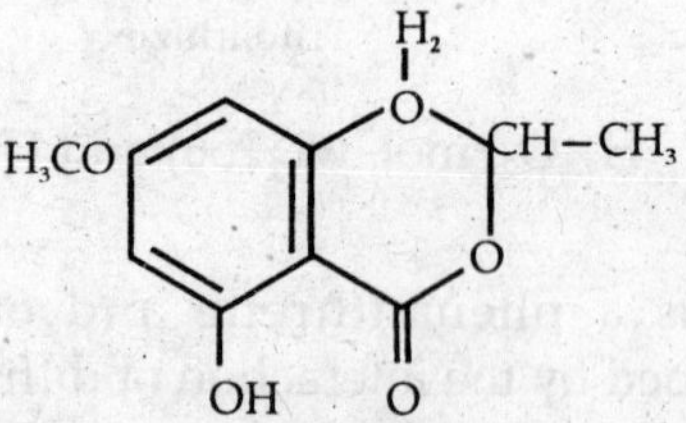

Isocoumarin

Medicarpin

Alfalfa *(Medicago sativa)* inoculated with a number of pathogens and non-pathogens of this host has been studied by Higgins and Millar. The antifungal compound was isolated from leaf diffusate solutions and has been identified as (–) demethyl-homopterocarpin and given the name, medicarpin.

Cicerin

Kunzru and Sinha reported the presence of a new phytoalexin associated with *As cochyta* blight of gram. The inhibitory principle is composed of two fluorescent phenols or phenolic compounds. Phytoalexin diffuses into inoculation droplets during the course of interaction between *Ascochyta rabie* and the pod tissues of *Cicer arietinum.* The name cicerin was tentatively suggested by them.

Rishitin

Muller and Boerger and Milller were the first to show that the potato tubers carrying the gene RI for late blight resistance responded when inoculated with avirulent race (race) of *Phytophthora infestans* by producing a phytoalexin that inhibited the development of a virulent race (race 1) applied a short time later. Two non-pathogens of potato (*C.fimbriata* and *Helminthosporium sativum)* also stimulated accumulation of rishitin in both kinds of tuber tissues. Tomiyama , in studies on potato tubers *(Solanum tuberosum)* infected with an incompatible race of *Phytophthora infestans,* isolated a new compound with phytoalexin-like properties. It is a bicyclic non-sesquiterpene alcohol (Katsui *et al.,* 1968) and has been given the name, rishitin.

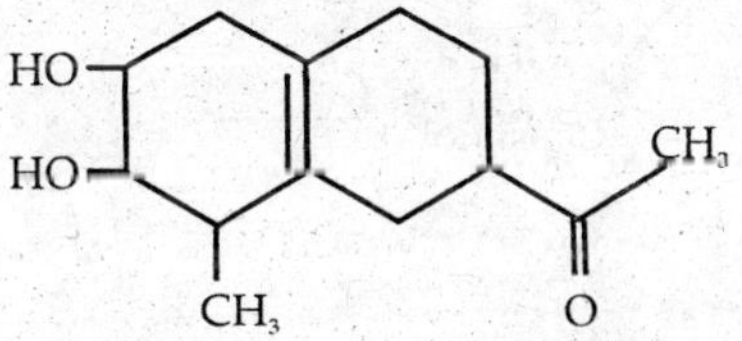

HO HO CH_3 CH_3 O

Rishitin

Gossypol

Bell has reported the amassing of several antifungal compounds due to the introduction of conidia of *Verticillium albo-atrum* and *Rhizopus nigricans,* heavy metal ions and various metabolic inhibitors into boll cavities of xylem vessels of the excised stems of two species of *Gossypium* and many cultivars of cotton. The compound has been identified as gossypol.

Other diseases in which there is proof of the production of phytoalexin-like substances include black spot of rose *(Diplocarpon rosae*), leaf spot of wheat caused by *Septoria tritici* & leaf disease of *Primula* caused by *Ramularia primulae* and *Cercosporella primulae*

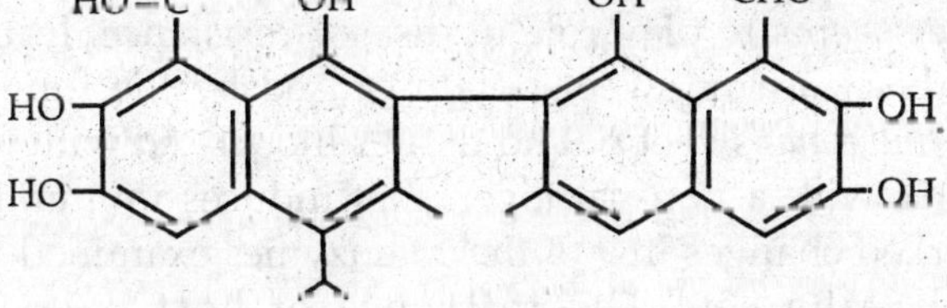

HO–C OH OH CHO HO HO OH OH

Gossypol

Capsidiol

Diffusates from pepper *(Capsicumfrutescens* L.) fruits grafted with a number of non-pathogenic fungi accumulate a sesquiterpene phytoalexin capsidiol in quantities that are sufficient to inhibit these fungi *in vitro.*

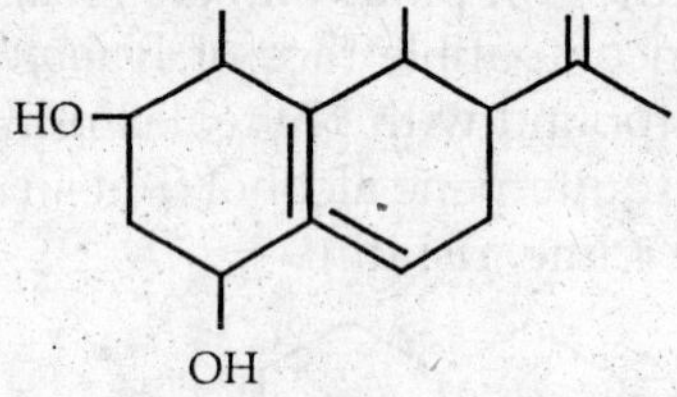

HO

OH

Capsidiol

Xanthotoxin

This phytoalexin was seperated from the top millimetre of parsnip root discs grafted with *C.fimbriata,* a non-pathogen of parsnip. Inoculation with other non. pathogens of parsnip also resulted in the production of Xanthotoxin.

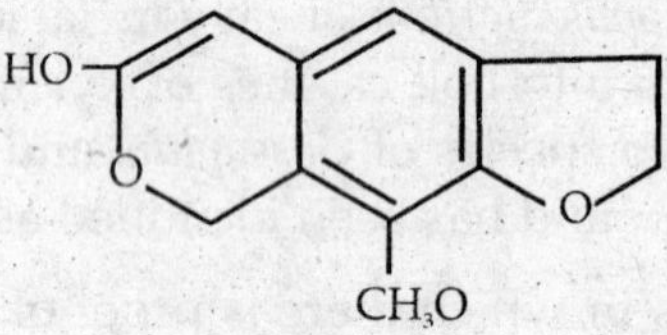

HO

O

O

CH_3O

Xanthotoxin

PROTECTION BY SYNTHESIS OF ENZYMES & PROTENS

In addition to the phenolic substances synthesized in plant tissues due to infection, induced production of proteins and enzymes seems to play role in disease resistance. In the case of the black rot of sweet potato caused by the fungus *Ceratocystis fimbriata,* sweet potato tissues inoculated with the pathogen and with a non-pathogen that induces immunity, showed marked changes in 9 of the 13 enzymes examined by Weber and Stahmann. The influence of host parasite interactions on proteins, enzymes and resistance has been discussed by Stahmann. Nucleic acid and protein changes in wheat leaf nuclei during rust infection were studied by Bhattacharya. There were decreases in histones and marked increases in RNA and acidic proteins as early as two days after inoculation in the nuclei of the mesophyll cells of a susceptible wheat infected with *Puccinia graminis tritici*.

In many incompatible host parasite combinations, peroxidase activity is often many times higher than in compatible ones. In other cases, however, peroxidase activity is greater in compatible reactions. It is, therefore, clear that the activation of peroxidase in the host and parasite complex need not always be accompanied by the incompatible reaction.

PROTECTION BY SYSNTHESIS OF SUBSTRATES WHICH RESIST PATHOGENIC ENZYMES

The resistance of green immature apple fruit to several fungi which can infect mature fruit has been attributed to the presence in the green fruit of a pectin-protein-polyvalent cation complex which is resistant to hydrolysis by fungal hydrolytic enzymes and decreases the production of pectin methyl esterase and polygalacturonase by the fungi.

An induced mechanism of resistance through increased resistance of the substrate to enzymes by the accumulation of polyvalent cations has been reported for *Rhizoctonia* infected bean tissues. In the resistant variety, calcium ions accumulate in and immediately around developing *Rhizoctonia* lesions, inhibit further tissue maceration by polygalacturonase, and therefore, play a role in confining the pathogen to lesions of a limited size. Fungal PME demethylate pectic substances near and in advance of the pathogen and result in the formation of pectate salts with polyvalent cations, particularly calcium, which in turn are resistant to hydrolysis by fungal polygalacturonase.

Disease resistance introduced through artificial application of auxins to infected plants has been reported for *Fusarium* wilt of tomato and late blight of potato. The resistance to disease in auxin-treated plants is accompanied by higher levels of insoluble pectates and lower levels of water soluble pectins than are found in untreated, susceptible plants.

The resistance of cider apples to wound infection by the brown rot fungus *Sclerotinia fructigena* seems might be due to the ability of phenolic oxidation products in these apples to inactivate extracellular pectinolytic enzymes produced by the fungus. The phenolics synthesized in the juice of the resistant variety do not inhibit the growth of the fungus, but do inhibit the activity of its polygalacturonase.

PROTECTION BY TOXIN DETOXIFICATION

Detoxification may work in resistance to vascular wilt diseases caused by *Fusarium oxysporum*. Its significance depends on the importance properly ascribed to toxins in these diseases. Lakshminarayanan reported that varieties of cotton resistant to *F. oxysporum* f. *vasinfectum* have relatively high quantities of cystine and that cystine reduces the toxicity of the complex formed when ferric iron is added to filtrates from the cultures of the parasite. When shoots are allowed to absorb cystine from solutions before the complex is transferred to the former, they are less damaged than are plants not pre-treated with cystine.

In the case of the toxin, piricularin, Tamari found that chlorogenic acid could detoxify the piricularin when applied to rice seedlings in equimolar concentrations. Tamari also showed that piricularin was detoxified with ferulic acid, which was oxidized by peroxidase, and its oxidized products were also shown to detoxify the piricularin. Those varieties of rice which have more chlorogenic or ferulic acid are more resistant to the blast disease.

The partially purified toxin produced by *Alternaria kikuchiana* is inhibited by various kinds of oxidizing substances, including dehydroascorbic acid. Mori (1962) showed that the resistant varieties of pear contain more dehydroascorbic acid and more ascorbic acid oxidase than the susceptible varieties.

PROTECTION BY ALTERING RESPIRATION

The relation of enhanced respiration to disease resistance and susceptibility is unclear. In some incompatible host pathogen combinations (resistance), the increase is initially greater than in compatible combinations (susceptibility), although the increase may be of greater total magnitude and longer in duration in susceptible reactions.

From observations, however, indicating that substances inhibiting respiration also diminshes resistance to disease in the host, it is suggested that increased respiration is often a necessary condition for the development of defense reaction in the host. A marked increase in oxygen uptake has been reported for the hypersensitive defense reactions. This is also accompanied by the accumulation and oxidation of phenolic compounds and by the activation of phenoloxidases.

PROTECTION BY HYPERSENSITIVITY

Hypersensitivity reaction or hypersensitivity works as a defense reaction in plants against pathogens. It happens only in incompatible host-parasite combinations situation. This is a significant and common form of reaction in plants. Usually, a small part of the plant tissue quickly dies after infection, resulting in a minor or local lesion or one as extensive as to cause the death of the plant. Hypersensitive reaction is believed to be produced by the disturbance of the balance between the oxidative and reductive processes in the cells resulting in excessive oxidation of phenolic substances and leading to the breakdown of cellular structures (Figure)

GENERAL NOTION OF PHYTONCIDES

In 1930, at the All-Union Conference of Zoologists in Kiev, USSR, Tokin showed the results of his studies on the antibiotic properties of plants. Continuing this work, he published a antibacterial substances formed by plants which he called phytoncides. This term was later used by other Russian workers to include the substances of plant origin active against other organisms, both animals and plants. Phytoncides, by definition, include phytoalexins; and pre-formed and postinfectionally formed substances which give resistance to plant pathogens, in addition to those which deter or destroy insect pests.

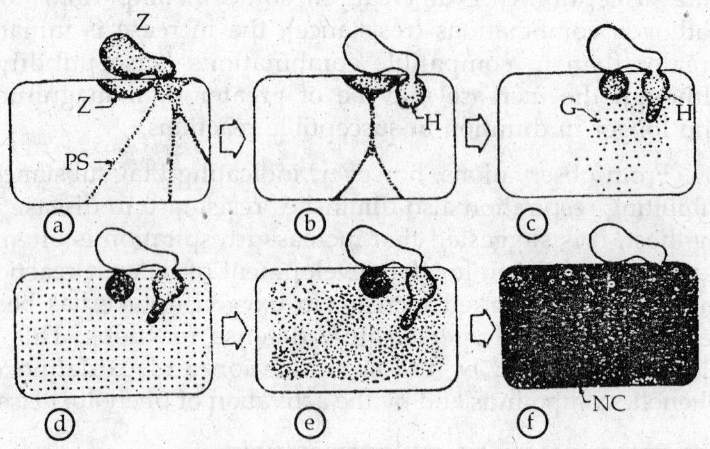

Figure. H Different stages in the development of necrotic defense reaction (hypersensitivity) in a cell of a very resistant potato variety infected by phytophthora infestans. N=nucleus, Z=zoospore, PS=protoplasmic stands, H=hypha, G=granular material, NC=necrotic cell.

Rubin and Artsikovskaya (1963) studied this aspect of phytoncides in detail. It is being felt by plant pathologists that the term is too vogue to be of much use and the individual categories of chemical substances known so far are adequately covered by more specific terms.

10

Transmission of Disease Causitive Organisms

Distribution implies the spread of a plant pathogen into new geographical areas and its establishment there. The spread of a plant pathogen within the general area in which it is established is termed dissemination or dispersal. Contamination -the movement of inoculum (even if only a few inches) to a susceptible plant-is a kind of dissemination. A knowledge of the methods by which pathogens are dispersed is usually essential before effective control measures can be devised. There are essentially two types of dissemination of plant pathogens:

A. Direct transmission.

B. Indirect transmission.

In direct transmission, the dispersal takes place along with the seeds and vegetative parts, while indirect transmission may be autonomous, that is, by wind, water, animals or human beings.

Autonomous dispersal by plant pathogens has been discussed by Muskett (1960), dispersal of inoculum by insects and other animals including man, has been discussed by Broadbent (1960) and liberation and dispersal of fungal spores have been ably discussed by Ingold (1971) in his book, *Fungal Spores.*

A. *Direct Transmission*

a. Germinative transmission

b. Vegetative transmission

c. Adherent transmission

B. *Indirect Transmission*

a. Autonomous transmission

b. Wind dispersal of pathogens-Anemochor

c. Water dispersal of pathogens-Hydrochory

d. Animal dispersal of pathogens-Zoochory

e. Human dispersal of pathogens-Anthropochory

DIRECT TRANSMISSION

Germinative Transmission

This takes place along with the seeds or propagules. Bean mosaic is transmitted through pollen grains and carried in the seeds. In the case of loose smut of wheat and barley caused by *Ustilago tritici* and *U. nuda,* respectively the seeds get infected without producing any symptoms in the parent carrier. The seeds carry the intraseminal mycelium. When such apparently healthy looking seeds are sown, the symptoms appear when the ears emerge. The pathogen develops systemically inside the plants. Alternaria leaf blight of wheat is both internally as well as externally seed borne (Prabhu and Prasada, 1966). The Helminthosporium blight of wheat and barley is seed borne. The seed borne infection is present in the form of mycelium within the seed including the endosperm and embryo. The fungus has been found to remain viable up to seven years on barley seed (Christensen, 1922). Bacterial blight of paddy caused by *Xanthomonas oryzae* is believed to be seed borne.

Vegetative Transmission

Transmission of a large number of plant pathogens takes place through the vegetative parts of plants used as seed such as tubers, cuttings, runners, grafts, etc.

Ring rot of potato caused by *Corynebacterium sepidonicum* is carried through infected potato tubers to new areas, as in the case of the pathogen *Pseudomonas solanacearum* responsible for brown rot of potatoes. A large number of viruses are transmitted through vegetatively propagated parts.

Dormant mycelium of *Phytophthora infestans* is carried through the potato tubers. Seed setts of sugar cane carry the pathogen, and when such apparently healthy looking vegetative parts are sown, the pathogen develops systemically. Whip smut of sugar cane caused by *Ustilago scitaminea* and red rot of sugar cane caused by *Colletotrichum falcatum* are 'seed' borne (*i.e.* through the diseased cane sets).

Adherent Transmission

In this case, the propagules of the pathogens are carried over the surface of the seeds or vegetatively propagated parts. Bunt of wheat caused by *Tilletia foetida* and *T. caries* is carried through the seeds externally, as in the case of covered smut of barley, where the smut spores attach themselves to the surface of the seeds at the time of threshing, There are many other fungi whose spores remain adherent, such as *Synchytrium endobioticum* responsible for potato wart and species of *Helminthosporium, Rhizoctollia solani* (black scurf) sclerotia on the tuber surface of potato and *Alternaria* which causes various diseases.

INDIRECT TRANSMISSION

Autonomous Transmission

Autonomous dispersal has been discussed in great detail by Muskett (1960). Plant pathogens can be spread to short distances by this method. Autonomous transmission takes place by active growth of hyphae or hyphal strands. It is characteristic of wood rotting fungi, such as *Armillaria, Fomes, Ganoderma, Polyporus* belonging to the Hymenomycetes which nigrate independently through the soil from plant to plant or even from field to field by the active growth of their hyphal strands.

Armillaria mellea rhizomorphs spread from one root to the other in forest trees. Hyphal mats or strands' grow out to short distances in the soil from the sclerotia of *Rhizoctonia solani* and *Sclerotium rolfsii*. The rate of such spread in *Phymatotrichum omnivorum*, responsible for Texas root rot of cotton crop is estimated at 5 to 30 feet per season and 2 to 8 feet per month in alfalfa crop. It may be emphasized that in *Armillaria mellea* the cause of the tree root disease, *Ophiobolus graminis*, the cause of the take all disease of wheat, *Phymatotrichum omnivorum*, the cause of root rot of cotton, the pathogens do not grow and ramify through the soil as do the nonspecialized types, but require a food base, such as the roots or crop debris of the host plants which they attack.

Dispersal of Pathogens by Winds

This type of spread is termed by Gaumann (1950) as 'Anemochory'. Some plant pathogenic bacteria are carried to short distances by the wind. The seeds of some angiospermic plant parasites are effectively carried by the wind. All fungal pathogens which produce spores or conidia on the surface are disseminated by the wind, such as downy mildews, powdery mildews, rusts, smuts, leaf spot causing pathogens, sooty molds, etc. The pathogen must produce numerous spores which are successfully liberated, dispersed and deposited in a viable condition on susceptible plants under conditions conducive to infection; in other words, production, liberation, dispersal and deposition of spores are involved. Dispersal through the air has been ably reviewed by Harald Schrodter (1960).

The fact that spores of fungi are disseminated by air currents has been known almost ever since the existence of spores themselves has been known. In 1929, Micheli published the results of epoch-making investigations on the reproduction of fungi by means of spores and showed that clouds of them might be liberated into the air.

Many fungus pathogens are remarkably well adapted to wind dissemination, for they produce and liberate into the air almost countless numbers of small and very light spores that are carried to long distances.

Propagative spores are produced in many different ways in sporangia, on simple or complex conidiophores, in or on ascocarps and basidiocarps, by budding or fusion, within pycnidia, perithecia or sori of various types, and in other ways, as described in books on mycology and plant pathology (Stakman and Harrar, 1957; Hughes, 1953; Hawker, 1960). Fungi show a great diversity in their methods of producing propagules and many different ways of more efficient production and liberation of spores seem to have been explored. Spore production per unit area of infected leaf tissue may be increased by close aggregation of sporophores, branched or whorled sporophores, production of spores in chains or clusters, by a more or less continuous production of spores within the sorus or by successive crops of spores. The number of spores produced can be astronomical. Powdery mildews may produce several thousand conidia per square centimeter of infected leaf surface, a fairly modest output compared with 10,00,000 or more some downy mildews. A single smut sorus may contain millions of spores, a heavily infected barberry bush is said to produce up to about 70,000 million aeciospores of *Puccinia graminis* at One time (and there may be several crops of spores in the spring), and a relatively small apothecium of *Sclerotinia* can produce about 30 million ascospores. The spores produced by large polypores, such as *Fornes* are even more numerous. Heavily rusted wheat fields produce so many spores that spore clouds may be seen rising during harvesting and implements too may be layered with a fine reddish deposit of spores.

The duration and periodicity of sporulation is as important as the number of spores. Some pathogens produce a more or Less simultaneous 'crop' of propagules followed by other crops, as in the case of the sporangiophores of *Peronospora*. The sporophores of others may bear several successive propagules, for example, those of *Phytophthora*, *Cercospora*, etc. The capacity to produce a steady stream of infective propagules over a prolonged period of time is likely to be advantageous to the pathogen. Some fungi show a diurnal periodicity in sporulation, as in nocturnal production of sporangia by some species of *Sclerospora.* This is probably chiefly due to temperature moisture effects, but might in some cases be related to periodicity in susceptibility of the host plant. Similar periodicity in spore discharge (Hirst, 1953) and in the germination and formation of appressoria (Yarwood, 1936a) has been reported.

Apart from spores, bits of mycelium and nematode cysts are also sometimes disseminated by wind. In India the molya disease of wheat and barley *(Heterodera major)* is believed to have been introduced into Haryana from Rajasthan.

According to Schrodter (1960), the problem of dissemination of plant diseases where propagules cannot move by themselves is more amenable to theoretical treatment than any other problem in phytopathology. Fungal spores in air behave as inert particles, perhaps similar to those in smoke, with terminal velocities ranging from about .05 to about 2.5 cm per second, the larger spores falling more rapidly than small ones.

Although fungus spores are generally more abundant in the air near the earth than at high altitudes, several investigators have encountered clouds of spores, numerous bacteria, and other minute objects several thousand feet above the earth. Urediospores of *Pucdnia graminis tritid* have been caught as high as 14,000 feet above infected grain fields, living spores of various fungi were caught from aeroplanes above the Carribean sea 600 miles from their nearest possible source and living spores of several common molds were caught in a spore trap released from the baloon Explorer II at 72,500 feet and set to close at 36,000 feet.

Long distance dissemination of rusts is a well established phenomenon. In the Indian subcontinent, rust spores are known to make big jumps *from* the source areas to the plains. (Mehta, .1940; Nagarajan and Singh, 1974, 1975).

Studies by Nagarajan and Singh (1973, 1974,1975) have shown that for South and Central India the spores are wind borne, rain deposited and are transported from an altitude of around 700 mb (3030 m). The uredospores get transported by upper winds from the Nilgiri and Palney hills and are then washed down over Central India by rain. Such a spread occurs to a distance of 600 km or more without infecting the fields in between.

Turbulence redistributes spores and affects the progressive dilution of spores with increasing. distance from their source. There is ample evidence that with normal wind and turbulence conditions, spores, especially small ones, can travel considerable distance. Those of *Phytophthora infestans* have been reported by various observers to travel distances of 200 m to more than 60 km depending upon the wind strength and degree of turbulence.

The effects of air currents on spore dispersal are complex and varied. A steady wind blowing fairly strongly tends to carry spores horizontally without allowing them to rise to any extent, except on encountering obstacles in their path. Slight breezes and irregular gusts blowing in different directions can cause confused air movements if convection currents are rising and this turbulence may carry spores to considerable heights, thus facilitating their dispersal. These irregular air movements are often more effective in bringing about multidirectional spread of pathogens than are strong sturdy winds. The whole complex problem of deposition in phytopathology has till now been neglected. A few studies have been concerned with the visible results of deposit processes rather than with the processes themselves. 'Warwood and Hazen (1942) made studies on how the conidia of *Erysiphe graminis* are oriented during deposition and thus touched an important question. Gregory (1951, 1952) concerned himself with deposition in a more extensive way and made valuable experimental observations on exactly defined surfaces.

Water Dispersal of Pathogens (Hydrochory)

Water as an agent of dissemination appears to be relatively less important as compared to wind. Although it sometimes facilitates dissemination in various ways, water does not furnish the motive power for long distance dissemination, except as streams and rivers. Flood and irrigation water may carry inoculum, especially that in or near the soil. It is clear that water separates and distributes spores in a micro environment. Spores that are extruded in gelatinous tendrils that may harden when dry, can be separated from each other and washed down trees and other kinds of plants during rain or heavy dews. The splashing and spattering of water during heavy rains may result in distributing inoculum to plant parts near the soil and may distribute bacteria and spores to different parts of the same plant or to neighbouring plants.

In some places, rain seems to be more important than insects in disseminating the fire blight bacteria, *Erwinia amylovora,* by its washing or spattering effect. The secondary spread from initial infection of *Xanthomonas campestris* causing black rot of cabbage and other crucifers, of X. *translucens* which attacks cereals and grasses and of bacterial blight of tomato, *Corynebacterium michiganense,* results largely from splashing during rains.

Xanthomonas malvacearum, the bacterium that causes angular leaf spot of cotton, can be spread by wind blown rain.

Surface water resulting from rains or irrigation is known to disseminate some pathogens and undoubtedly a large number of spores. The bacterium causing angular leaf spot or black arm of cotton is spread largely by irrigation water or surface water in Arizona, USA and Uganda. *Plasmodiophora brassicae,* the organism causing club root of crucifers, is spread by surface water, and *Phytophthora infestans,* causing late blight of potatoes, has been observed to spread from plant to plant in dry areas by means of irrigation waters.

Colletotrichum coffeanum which produce spores within the acervuli in the maturing, bark of coffee twigs and on diseased berries is'disseminated by water.

Water, then, is an important agent of local dissemination, but it is not as important as wind for long distance dissemination.

Large drops of water in the form of rain or overhead irrigation, falling at their terminal velocity, are probably most active in splash dispersal, much more so than small drops. In a splash dispersal, the spores are carried and often deposited on the plant surface in a droplet of water which could provide the moisture required for their germination especially in humid weather. Splash dispersal, although usually considered a mechanism for liberating and dispersing slimy spores, may also play a part in the dispersal of roughened spores which adhere strongly to each other or to plant surfaces. An example of this is provided by *Hemileia vastatrix* (coffee rust) in which, according to Nutman (1960), the roughened uredospores cling together and are broken up by water. They then float singly to the surface of the water drop, from which they are splash dispersed by further rain drops. The cells of many plant pathogenic bacteria and the slimy spores of such important genera of fungal plant pathogens as *Septoria, Fusarium* and *Colletotrichum* are splash dispersed and it is surprising that relatively little information on splash dispersal is available.

The eelworm disease of potato and *Piper longum* have been known to be carried through surface water from field to field. The spread of infection in the red rot of sugar cane caused by *Colletotrichum falcatum* and the dissemination of the sclerotial bodies of *Sclerotium rolfsi* through soil are chiefly brought about by irrigation water. Water is also an important agency of dissemination for the seeds of flowering parasites such as *Orobanche*, dodder *(Cuscuta)* and *Striga* through the soil. Short distance transmission of semi-aquatic patho gens from one leaf to another of' an adjacent-plant occurs through splashing rains accompanied by wind such as for *Phytophthora arecae, P. parasitica, P. infestans* and other downy mildew fungi. These fungi are well adapted to wind as well as water dissemination due to the formation of deciduous conidia capable of germinating either by zoospores or germ tubes and thus develop into dangerous pathogens capable of widespread damage within a short period.

Diffusion by Insects, Mites and Nematodes

Insects

Since the discovery by Waite (1891) that bees and wasps can transmit bacteria causing fire blight of apple and pear, *Erwinia amylovora,* much information has been accumulated on the role of insects and other small animals in the dissemination of plant pathogenic viruses, bacteria and fungi. Many small animals undoubtedly disseminate some pathogens occasionally and incidently but insects fall into a special category.

Insects not only transmit and disseminate many plant pathogens but are sometimes agents of inoculation also. Moreover, some pathogens-multiply within insects and some are present in them during periods when host plants are dead or dormant.

The various interesting relationships between micro-organisms including plant pathogens and insects have been discussed in detail by Leach (1940) and Carter (1962). Shorter accounts have been given by Broadbent (1960), Gaumann (1950) and Austwick (1957) on fungi, by Bawden (1964) on viruses and by Madelin (1966, 1968) on entomogenous fungi.

The largest number of insect vectors have sucking mouth parts which belong to the Hemoptera. Within this order the largest number of species' which are known to transmit some viruses or the other is contained within two families, Aphidae (aphids) and Cicadcllidae (leaf hoppers). A few belong to the Aleyroideae (white flies), the Cercopidae (Spittle insects), the Coccidae (mealy bugs) and the Membracidae (tree hoppers) families. Thrips transmit the spotted-wilt virus. some viruses are transmitted by both chewing and sucking insects, while a few are transmitted only by the former. Squash mosaic, cowpea mosaic and turnip yellow mosaic viruses are transmitted by beetles (Coleoptera) while the~ last is transmitted also by grass hoppers (Orthoptera) and earwigs (Dermaptera).

It is of interest that practically all vectors of the so-called yellows group of viruses are leaf hoppers while most vectors of the mosaic group are aphids. In other words, those viruses which are more readily transmitted mechanically are aphid borne and those which do not yield to such transfers are leaf hopper borne.

More than fifty species of aphids are known to transmit the viruses of plants. Some species can transmit only a few kinds and others transmit many; *Myzus persicae,* the green peach aphid, alone is the vector of fifty different viruses.

Some of the diseases transmitted by aphids are Katte or Marble disease of cardamom, mosaic and grassy shoot of sugar cane, mosaic streak of wheat, maize mosaic, barley mosaic, bean mosaic, cowpea mosaic, bunchy top of bananas, banana mosaic, papaya mosaic, potato necrosis (viral), potato leaf roll, tomato mosaic and chilli mosaic.

Some of the diseases transmitted by white fly are tobacco leaf curl, mung *(Phaseolus aureus),* yellow mosaic, double bean *(Phaseolus lunatus)* yellow mosaic and yellow mosaic of *Dolichos lablab.* Tungro of rice is transmitted by leaf hopper vectors, while bottlegourd mosaic is transmitted by red pumpkin beetle.

Three important questions arise with respect to the transmission of viruses by aphids:

a. How long must the insect feed on infected plants before it can transmit the virus to healthy plants?

b. Is an incubation period in the insect necessary before the virus can be transmitted?

c. How long do the insects remain infective after having acquired the virus ?

Some viruses are quickly acquired and lost, others are acquired more slowly and are I quite persistent. Some require an incubation period and others do not. In general, the viruses that require no latent or incubation period in the insects are relatively nonpersistent and those that require one or more are persistent. Insects are important in the dissemination inoculation and persistence of certain bacterial pathogens of plants.

Erwinia tracheiphila, the cucurbit wilt organism is completely dependent on cucumber beetles for its spread. *Xanthomonas stewarti,* the corn wilt pathogen, overwinters inside the corn flea beetle *(Chaetocnema pulicaria).* The cucumber beetles, *Diabrotica vittata* and *D, duodecimpunctata,* chew the plant parts affected by *E. tracheiphila* and carry the bacteria inside the body where they overwinter, passing on the host in the next season. It is believed that the relationship between the insects is symbiotic.

Insects are important agents of dissemination and inoculation of *Erwinia carotovora,* causing black leg of potato. The pathogen survives in the seed tubers, which can account for primary infection in the spring. It was originally thought that the organism did not survive in the soil and there was no secondary spread in the field. Circumstantial evidence later indicated that there must be a secondary spread, but the agents remained unknown until Leach in 1926 proved that the seed corn maggot, *Hylemya cilicrura,* is a very important agent of effective dissemination and inoculation.

The seed corn maggot and closely related species are the pests of many kinds of plants such as corn, cabbage, beets, beans, pear, radish, onion, tomato, etc. These insects and certain others contribute to the development of soft rot caused by *Erwinia carotovora*'in some of these crops in much the same way as has been described for black leg.

Erwinia amylovora the cause of fire blight of apples, pears and certain other members of the Rosaceae, was the first plant disease known to be caused by bacteria and transmitted by flies and ants. *Xanthomonas citri* which causes the well-known citrus canker is carried by leaf miner from diseased to healthy plants.

Xanthomonas stewartii and *X. vasculorum* which cause wilt of corn and gummosis of sugar cane, respectively, are transmitted by insects and can persist inside the vectors. *X. stewartii* is also transmitted on seed and *X. vasculorum* on sugar cane setts used for planting. The corn wilt organism, however, is carried from plant to plant in the fields by flea beetles. *Chaerocnema Plilicaria* and C. *denticlilata* by *Diabrotica duodecimpunctata* mentioned in the discussion on cucurbit wilt, and probably by other insects as well. Moreover, the bacteria can live during the winter inside the corn flea beetle, C. *pulicaria,* and can live for some time in the adult *Diabrotica* beetles.

The big-vein disease of lettuce has been :;known since the 19308. The disease causing agent was not identified at that time but was known to be soil-borne. In the 1950s the chytridiomycetous fungus, *Olpidium brassicae* was found to be intimately associated with this disease. Finally, in 1961, Camphell showed .that the causal agent could be transmitted by grafting in the absence of *Olpidium.* Thus, this disease is now known to be caused by a virus. Later work by Campbell has shown that this virus is transmitted by the zoospores of *Olpidium* and survives within the resting spores of this fungus. *Olpidium* also transmits other viruses that are soil borne. They are also known to be transmitted by various chytridiomycetes.

According to Austwick (1957), about 66 species of fungi causing plant diseases are transmitted by more than a 100 species of insects belonging to at least six orders. Transmission occurs in various ways with or without injury to the plant.

Spores of *Ustilago violaceoe* (another smut of Caryophyllaceae) and *Botrytis anlhophila* (anther mold of clover) are thus carried from the infected anthers to, uninfected flowers by pollinating insects. The conidia of *Claviceps* are spread from infected ovaries to healthy ones by insects which feed on the sugary honey dew on which the conidia are produced. Other fungi infecting the shoots and roots of plants are likely to be fortuitously disseminated by insects which move or are carried from plant to plant. Richardson and Saunders (1968), for example, attribute the partial control of potato blight by aphicide sprays to the destruction of aphids which carry mycelial fragments and sporangia of *Phylophthora infestans* on their bodies and leg bristles, and so spread the pathogen over short distances.

More frequently, however, transportation is accompanied by injury to the plant-the insect carries the fungus and places it in the wounds, thus facilitating infection. The injury may be caused by mechanical breaking, by other animals or by the insect itself. The spores of *Ceratocystis fagacearum* (oak wilt) are carried by various insects, chiefly nitulid and scolytid beetles, to existing wounds caused by breaking, wood peckers, squirrels and other causes. Mycelial mat. of the fungus develops beneath the bark, the crack is open and gives off a characteristic odour which apparently attracts insects. These feed on the fungus and also carry small spores which fertilize the receptive hyphae and so lead to the development of perithecia of the fungus. In some cases the insect or the fungus is adapted to insect transmission. Female Siricid (Sirex) wood wasps carry hyphae and conidia of *Stereum sanguinolentum* (heart rot of conifers) within the intersegmental, sacs at the anterior end of the long ovipositor. The eggs are contaminated with the fungus during laying and are deposited within the wood of the tree and the fungus is reported to precede the larva as the latter bores through the wood. The fungus is contained in the hypoplural organ modifications of the larval integument. *Ceratostomella ulmi* (Dutch elm disease) grows within the egg galleries of bark beetles *(Scolytus)* and contaminates emerging adults which also inject the slimy spores and thus spread the fungus to adjacent healthy trees. More specific are the diseases (stigmatomycoses) in which the fungus is injected into plant tissue by the insect vector. Several parasitic yeasts, notably *Nematospora,* are transmitted in this way, the acicular ascospores being sucked up the feeding stylet of stainer bugs *(Dysdercus, Nezara)* whence they lodge in the stylet pouches and are injected into young cotton bolls causing internal boll disease due to *Nematospora gossypi* (Frazer, 1944), bean pods *(N. phaseoli,* yeast spob) or coffee berries *(N. coryli, N. gossypi)*. It is thought that *Nematospora* is almost entirely dependent on plant bugs for dissemination and entry into the plant. In the case of chestnut blight caused by *Endothia parasitica,* the insects responsible for the disease are European bark beetle *(Scolytus multistriatus)* and the native elm bark beetle *(Hylurgopinus rufipes)*.

Several insects attack the fruits of plants and fruit rotting fungi often enter through the wounds. The spores of *Aspergillus niger,* which causes smut of fig fruits, are reported to be I introduced by certain beetles and fruit flies, and *Sclerotinia* 'sp. (brown rot of various fruits) I can enter through feediug and oviposition wounds made by insects. Some fungi causing I stem rot (e.g. *Physalospora tucumanensis,* red rot of sugarcane) also invade through insect I wounds and it is suspected that some leaf attacking fungi behave in a similar way. *Pycnostysanus azaleae* (bud blast of *Rhododendrons)* is associated with the leaf hopper *Graphocephala coccinea* and the disease is very slight in the absence of the insect (Howell and Wood, 1962). Little is known of dissemination of the root pathogen by insects but it seem likely that this occurs in the soil and that insects feeding on roots also introduce fungi into them. The original insect damage may thus be greatly aggravated by subsequent fungal invasion. Insects which feed on cotton plants have .been found to contain *F. oxysporum* f. sp. *vasinfectum* (cotton wilt), the fungus having been recovered in a viable condition from faecal pellets of various grass hoppers. It seems possible that such insects disseminate this normally slowly spreading fungus which might explain the occasional appearance of the disease in hitherto disease-free fields or parts of fields, but other explanations are possible, and the relationship between the soil-borne pathogens and insects needs further investigation.

Insects which act as diploidizing agents for some heterothallic fungi, transfer fertilizing spores (spermatia, oidia) between compatible strains and so bring about the formation of the perfect state. The best-known example of this diploidization is that of receptive hyphae of rusts by insect-borne spermatia, but similar fertilization occurs in some Ascomycetes and other Basidiomycetes. This phenomenon has obvious similarities with insect pollination in higher plants.

Mites

These, although wingless, transmit some viruses (reversion disease of currants) which are perhaps carried by winged insects or wind. Transmission of fungal pathogens by mites has not been extensively studied.

Nematodes

Some viruses as well as some bacterial and fungal plant pathogens are transmitted by nematodes. *Corynebacterium fasdans,* leafy gall of various herbaceous plants; is carried by ecto- I parasitic nematodes *(Aphelenchoides)* and typical cauliflower symptoms develop only when both nematode and bacterium are inoculated into strawberry plants (Cross and Pitcher, 1952).

Yellow ear rot of wheat caused by *Corynebacterium tritici* is dispersed by the ear cockle nematode *(Anguina tritici).*

There is also evidence that some root infecting pathogens, particularly fungi,. may enter I roots through wounds made by nematodes and this may perhaps be of more frequent occurrence then is at present realized. Cotton wilt caused by *Fusarium oxysporum* f. sp. *vasinfectum* is an example of this, the pathogen being associated with root knot nematodes and probably others (Smith, 1953). The nematodes may also affect the physiology of the root, making it a more suitable medium for the fungus (Powell and Nusbaum,-1960; Powell. 1963). The reverse has also been reported; the fungus *(Pyrenochaeta lycopersici)* causing tomato brown root rot reduced the invasion of tomato roots by the potato root eelworm *Heterodera rostochiensis* (James, 1968). Bacterial wilt of carnations, caused by *Pseudomonas caryophylli,* is reported to be more severe in the presence of root knot nematodes which probably facilitate the entry of the bacteria (Stewart and Schindler, 1956). This has been discussed by Leach (1940) and by Pitcher (I965), Fielding (1959) Christie and Perry (1959).

In 1958, Hewilt *et al.* demonstrated that the grapevine fan leaf virus is disseminated by the nematode *Xiphinemaindex.* Both polyhedral and tubular viruses have since been shown to be nematode transmitted, the former by species of *Xiphinema* and *Longidorus* and the latter by *Trichodorus* spp.

Human Dispersal (Anthropochorus)

Man is, to a large extent, responsible for the dissemination of plant pathogens which he does in two ways-through his person and through the objects he transports from one place to another.

Local dissemination of pathogens on man's person is illustrated by the angular leaf spot disease of cucumber caused by the bacterium *Pseudomonas lachrymans.* When cucumbers a grown for pickles, the fruits are picked. When a person comes in contact with or picks an infected plant, his hands become contaminated, particularly when the plants are wet with dew. The pathogen is thus spread down the row of plants.

Another pathogen spread by man is *Corynebacterium michiganense,* the cause of bacterials canker on tomatoes. This pathogen is seed borne, and when it develops on tomato seedlings to be used, as transplants, the workers handling the transplants may contact the bacterium on their hands. Since the handling of transplants often causes small injuries to the plant, the worker placing the transplants into the ground can easily inoculate healthy plants. In California, USA, this disease has become a rarity since most tomato fields are no longer planted with transplants but are direct seeded.

The viruses that are spread by animals can also be spread by man. Tobacco mosaic virus is known to survive in dry tobacco leaves. Hence, workers who handle tobacco plants in the field are advised not to smoke while they are working and to thoroughly wash their hands before entering the field. The long distance dissemination of plant pathogens by man is usually accomplished by the transport of infected plant parts, such as seeds, nursery stock or timber products. The cause of some of the worst plant disease epidemics is the introduction of a pathogen into a new area. In many cases the plants growing in these areas are very susceptible since they do not develop resistance through evolution in the presence of the pathogen.

Table 9.1

Examples of Plant Diseases which have been Introduced in India from other Countries

Disease	Original home	Year of introduction
Leaf rust of coffee caused by *Hemileia vastatrix*	Sri Lanka	1879
Late blight of potato caused by *Phytophthora infestans*	England	1883
Flag smut of wheat caused by *Urocystis tritici*	Australia	906
Downy mi ldew of grapes caused by *Plasmopara viticola*	Europe	1910
Downy mildew of cucurbits caused by *Erysiphe cichoracearum*	Sri Lanka	1910
Downy mildew of maize caused by *Sclerospora phillipinensis*	Jawa	1912
Black rot of crucifers caused by *Xanthomonas campestris*	Jawa	1929
Powdery mildew of rubber caused by *Oidium heveo*	Malaya	1938
Black shank of tobacco caused by *Phytophthora nicotwnae*	Holland	1938
Crowngall of apples and pears caused by *Agrobacterium tumefaciens*		*1943*
Bunchy top of banana (viral)	Sri Lanka	1940
Hairy root of apple (viral)	England	1940
Bacterial leaf streak of chillies		1948
Onion smut caused by Urocystis cepulae	Europe	1958
Bacterial blight 'of paddy caused by *Xanthomonas oryzae*	Philippines	*1959*
Golden nematode of potato caused by *Heterodera rostochiensis*	Europe	1961

Table 9.2

Examples of Plant Diseases Introduced in some other Countries

Disease	Introduced home	Original home	year of introduction
latc blight of potato caused by *Phytophthora infestans*	*Europe*	S. America	1830
Powdery mildew of grapes caused by *Uncinula necator*	England	N. America	1845
Blister rust of pine caused by *Cronartium ribicola*	USA	Europe	1910
Downy mildew of vine caused by *Plasmopara viticola*	France	USA	1878
American Gooseberry mildew caused by *Sphaerotheca mors-uvae*	England	N. America	1899
Chestnut blight caused by *Endothia parasitica*	USA	Asia	1904
Cittus canker	USA	Asia	1907
Paddy blast caused by *pyricularia oryzae*	India	Sri Lanka	1918
Striga asiatica	N. America	Africa or Asia	
Rust of coffee caused by *Hemileia vastatrix*	Brazil	Africa and Asia	1970
Dutch elm discase caused by *Ceratosomella ulmi*	USA	Holland	1928 30
Fire blight caused by *Erwinia amylovora*	New Zealand	N. America	1919
Golden nematode cause by *Heterodera rostochiensis*	USA. Mexico, Peru, India	Europe USA	1881

Man has repeatedly carried pathogens to areas which otherwise would have remained free of them. Almost all countries have suffered to such an extent from pathogens introduced by man that the Food and Agricultural Organization of the USA and several international committees of plant pathologists are promoting studies and regulations designed to reduce the danger.

Man is an agent of transportation of pathogens from continent to continent, from country to country and even from plant to plant on his own farm. Eternal vigilance is necessary to protect economic plants against the pathogens distributed by man.

11

Forecasting Techniques

The biggest aim of the science of plant pathology is to have a complete understanding of plant disease-host, pathogen, environment interactions-so that it becomes possible to control disease economically. The best and most economical way to control disease is usually to get a thorough knowledge of the disease epidemiology. We cannot effectively control certain diseases without some previous indication as to how they will behave in the coming period, that is, without some knowledge as to where and when they will occur, which will enable preventive action to be taken. Unless we know where and when the sprays or dusts, etc. are to be applied then there is no use. For certain diseases, no effective or practicable chemical control measures are available. Forecasting gives us an opportunity to revise our plans in order to avoid or reduce losses.

The practical results of forecasting have been incorporated in the definition proposed by Miller and O'Brien who state: "Forecasting involves all the activity in ascertaining and notifying the farmers of a community that conditions are favourable for certain diseases, that application of control measures will result in economic gain, or on the other hand, and just as important that the amount of disease expected is unlikely to be enough to justify the expenditure of time, energy and money for control: Miller and O'Brien thus include a warning service for plant disease prevention in their definition of forecasting.

Forecasting the incidence of a plant disease must take into account not only the disease itself but the weather also, which is more or less predictable. Disease forecasting is made more reliable if the reasons for a particular disease-developing under certain conditions and not under others are known. Experimental investigation is must to show at exactly what stage in the development of a disease any single factor is and to show to what extent and how the influence of anyone condition is modified by different combinations. Experience with late blight for example, shows definitely that the relative importance of each separate weather element varies with the intensity and fluctuation of other conditions.

Holland pioneered the development of forecasting and spray warning services for the control of late blight fungus. Van Everdingen analyzed the combined effect of several weather conditions on the development of *phytophthora infestans* and evolved four rules, known as the "Dutch rules" upon which the appearance of blight was observed to depend. They are:

(a) dew during at least 4 hours at night, that is, night temperature below the dew point for this length of time,

(b) a minimum temperature of 10°C (50°F) or above,

(c) a mean cloudiness on the next day of 0.8 or more,

(d) measurable rainfall during the next 24 hours (at least 0.1 mm). When all these four conditions were obtained in Holland, control measures were immediately recommended.

In 1948-1949 a method was tried to find which would make it possible to give warnings without using the special station. It was shown that in many cases the stations gave only a local and not a regional warning. Farmers often asked the institute to predict a critical period for the outbreak of the potato blight. Although it was possible to provide reliable warnings based on relative humidity and minimum temperature observation at the 40 cm level, normal synoptic weather observations from the regular weather stations were utilized to forecast blight outbreaks. This method had two important advantages. The first one was that all observations needed could be taken from the weather maps and from the synoptic stations which sent their observations to the Meteorological Institute every hour, and second, the warning was regional, and hence, more useful from the farmers point of view. In some cases it was possible to predict a period in which the weather would fulfil the conditions of temperature and relative humidity so that the farmers had one or two days time to spray their potato crop before the predicted outbreak of blight. In many cases where critical periods were predicted locally, the prediction was followed by some rainy days when spraying of course was not effective.

The basis of such predictions is a comparison of numerous observations from 2.20 m and 40 cm levels. The criteria are:

(a) two consecutive days with a mean-relative humidity of 82% or more (further investigation showed that potato blight also broke out when in one of the two days the mean relative humidity descended to 79%), the mean being taken from observations at 8,14 and 19 hours and

(b) for one of the days at least, the minimum temperature must be 10°C or more.

During approximately the past half century English workers have experimentally tested the Dutch rules for late blight forecasting. Beaumont and Hodson (1930) added the fifth rule to those formulated by the Dutch to apply under conditions in England, particularly in Devon and Cornwall. The additional fifth rule called for periods of not less than two days in which the relative humidity (at 3.00 p.m.) was higher than 75%.

In 1937, Beaumont and Stanlland reduced these five rules to two, since equally good results were obtained from their use, viz. (a) minimum temperature of 10°C or over and (b) relative humidity not over 75% for at least two days. This modification of the Dutch rules was adopted for future use.

The need for a warning service for the late blight of potato and tomato in the USA was evident long before a cooperative forecasting project was established. Moreover, a considerable amount of experience of various groups had already shown that the appearance and spread of late blight and some related fungi could be successfully anticipated, provided that means for watching and reporting progress were available. The event that finally led to the establishment of the regional forecasting project was the severe, widespread and unexpected epidemic of late blight of tomato in 1946. The project was organized to include a warning service and contributory research for three downy mildew diseases, late blight of potato and tomato, blue mold *(Peronospora tabacina)* of tobacco and downy mildew *(Pseudoperonospora cubensis)* of cucurbits. In the whole of North America all the way into Canada, the seasonal programme of these crops is particularly well adapted to observations and predictions on a widespread regional basis. Canadian pathologists cooperate in the warning service.

The appearance and spread of late blight of tomato and potato blue mold and downy mildew of cucurbits are promptly reported to the Plant Disease Epidemics Unit at Beltsville, Maryland. The reports are compiled into a warning letter sent twice a week during the growing season to the pathologists in the various states, Canadian provinces and other authorized recipients. By watching the occurrence and severity of disease in areas with an earlier growing season and applying their own previous experience, together with the weather forecasts, the key pathologists are able to judge rather accurately whether or not the local crops are likely to be affected, and if so, how soon and how seriously, and can make their forecasts accordingly. Manufacturers of control chemicals and equipment also use the warning letters as a guide to allocate their products.

a. The disease is one which causes economically significant damage in terms of quantity or quality in the area concerned.

b. The disease is variable in seasonal impact-time of onset, speed of build up, ultimate level and an appreciable part of this variation can be attributed to weather factors acting directly or indirectly.

c. Control measures, whether curative or preventive, are available and can be operated at economically acceptable cost.

d. Information is available from laboratory investigations or otherwise on the nature of the weather dependence of the disease.

Forecasting and warning systems involve considerable expense and should be economically Justifiable, as in the following cases:

a. For destructive diseases of important cash or food crops.

b. For diseases which can be reliably predicted early enough for effective control measures to be applied.

c. For diseases against which economical control measures are available.

Weather Conditions During the Crop Season

(a) The metcorological factors which form the basis of plant disease forecasting are temperature and moisture.

(b) Epiphytotics develop when favourable combinations of these factors occur. Laboratory studies of the combination of factors are useful, but these cannot be completely depended upon as detailed observations for a number of years in the field are necessary before any accurate disease forecasting system can be formulated. Forecasting of late blight of potato caused by *Phytophthora infestans in Europe,* Britain, the USA and other parts of the world call be cited as a good example where the combinations of weather conditions for blight epidemic will be different for different climatic and agricultural conditions. The spread or the 'Tikka' disease of groundnut caused by *Mycosphaerclla berkeleyvii* and M. *arachidicola* (C. *arachidicola)* in the USA is favoured by diurnal periods of 10 hours or more with relative humidity at or above 95 % and with temperatures above or about 21°C. This might be utilized in predicting outbreaks of the disease (Jensen and Boyle, 1965, 1966).. The percentage infection. of the, two youngest leaves and a bud and the duration of sunshine helps to predict blister blight of tea in Sri Lanka (Kerr and Rodrigs, 1967).

Amount of Disease in the Young Crop

Information on the extent of disease in the young crop has been successfully utilized in forecasting wheat leaf rust in Oklahoma, USA. The amount of infection at the beginning of April largely determines the subsequent development of the disease.

Amount of Inoculum in Air, Soil or Planting Material

Plant disease forecasting can be based on the amount of inoculum available in the air. The number of spores of a pathogen can be determined by various air trap devices set up in fields or orchards. Some diseases are blown down from long distances and in such cases the study of airospora helps in predicting the outbreak of the diseases. In the same way, the outbreak of soil-borne diseases depends largely on the distribution and amount of inoculum in the soil. Methods have been devised to assess the population of the pathogen in the soil.

These studies give an idea of the likely intensity of the diseases. Similarly, the degree of contamination or infection of the planting material can be taken to be a useful condition of the likely incidence of the disease. The best examples of the amount of inoculum in air and its correlation to the incidence of the disease are apple and pear scab. In some areas seasonal inoculum originates from conidia produced on infected shoots but more often from ascospores liberated from the perithecia in the leave on the ground. In the latter case the outbreaks are likely to be associated with a discharge of ascospores in relation to the development of foliage. Forecasting can thus be based on this factor. In Germany, the perithecia mature through the winter and the temperature after the first of March is said to be critical to ascospore discharge. Moisture also plays a significant part in this. Ascospore discharge usually occurs a week or so earlier on scabbed leaves inoculated in the laboratory than in the orchard. This is then utilized for giving a warning to the growers of the imminent danger of scab.

In India the plant disease forecasting service is in its infant stage. For forecasting rust epidemics in the last few years a number of rain samplers have been set in stations where facilities exist for their handling. The rain analysis gives precise data on the coming of the inoculum in a particular locality. Based on this information as well as data collected by survey teams from time to time, a large number of wind trajectories have been prepared. The air-borne inoculum and its deposition from the southern hills has been traced by wind trajectories and weather satellite-cloud images. If ground level conditions are favourable for infection the early arrival of inoculum can lead to severe rust epidemics.

In the case of brown stripe downy mildew caused by *Sclerophthora rayssiae* var. *zeae,* payak have observed that the disease is serious in areas in India where average rainfall is 100-200 cm or more. In areas of low average rainfall the increase in disease seriousness may occur only if there is a continuous spell of heavy showers and cloudiness for few days accompanied by lowered temperature (25°C or less).

Brown leaf spot of rice is caused by *Helminthosprium oryzae.* Certain abnormal trends have always been associated with a severe outbreak of *Helminthosporium* disease. For instance according: to 'Sundraman, usually heavy rains at the time of ear formation followed by following and ill drained conditions had favoured the epiphytotic of 1918-19 in the deltaic tracts of the Krishna and Godavari.

In the case of bacterial blight of paddy, our knowledge of the relationship of the environment to the disease appears to be based on field observations rather than on studies under controlled conditions. The bacterial blight of paddy develops readily above 25° C and symptoms may not appear below 20°C. Goto have reported that a combination of rainy weather, strong winds and a temperature of 22-26°C favour the outbreak of the disease.

Ramalingam has discussed the results obtained in Andhra Pradesh in the development of rice blast. He showed that a temperature range of 25-27°C, relative humidity of 86-90%, mild winds and darkness (midnight to sunrise) were critical for the development of epiphytotic. Rice crop is most susceptible to blast at three stages of crop growth, viz. seedling stage beyond 21 days, post transplanting stage, that is, 25-50th day after transplanting stages coincide with favourable environmental factors, blast outbreaks occur.

Susceptible stages of the crop, as indicated above, blast outbreak is expected. Thus, with the help of meteorological data, the outbreak of blast can be forecast (Padmanabhan, 1965). Govindswamy (1970) stressed the need for recording the number of low temperature days (20°C and less) in forecasting blast of rice. There is considerable scope for work in the study of epidemiology and forecasting outbreaks of plant diseases in India and there is a need for close cooperation between agricultural meteorologists and plant pathologists.

12

Control of Diseases

The main methods of controlling plant disease are: (a) avoidance, (b) exclusion, (c) eradication, (d) protection and (e) immunization.

In order to remove all the obstructions for maximum production of food, it is important to control the spread of plant pathogens by cultural methods, chemical methods and by breeding disease resistant varieties.

CULTURAL METHODS OF DISEASE CONTROL

Stevens discussed the cultural methods of disease control. According to Stevens, these methods involve agricultural cropping, harvesting and storage, tillage, crop rotation, soil management, growing of resistant varieties, planning of land use, and other practices.

1. Avoidance of the Pathogen

Many plant diseases can be totally avoided by a properly selecting of the land or field, time of sowing, selection of varieties, seed and planting stock and by improving of cultural practices. The aim of these measures is to enable the host to avoid contact with the pathogen or the susceptible stage of the plant and the favourable conditions for the pathogen should not coincide.

a. Selection of Sowing Area

Many fungal and bacterial diseases can be seen more in wet areas than in dry areas. The crops which are prove to these diseases, if grown in wet areas, are very likely to be affected by plant pathogens. *Tolyposporium penicillariae,* the pathogen responsible for the smut disease of bajra and the ergot of bajra caused by *Claviceps microcephala,* are more severe in wet areas in regions where rain occurs for long durations during the flowering stage of the crop.

b. Selection of Field

The selection of a suitable area or field for growing of crops is very important from the point of view of better results as well as protection of the crop from the clutches of pathogens. In the case of many soil borne pathogens, if the same field is chosen for cultivating a specific crop, there is every chances of the disease appearing in severe form due to the build up of the inoculum potential of the pathogen. Hence, it is advised not to grow the same crop in the same field, where in the previous year, it was infected with a pathogen capable of surviving in the soil. The examples are wilt of arhar, green ear disease of bajra, stem, powdery mildew of wheat, smut of bajra, ergot of bajra, ear cockle of wheat and bacterial wilt of solanaceous -plants.

The drainage conditions of the field are also important. Low lying, waterlogged fields will produces such diseases as red rot of sugarcane and downy mildew of bajra.

In the case of fruit orchards, the selection of a suitable field is very important. If apple trees are planted in areas where oak trees were previously grown, there are chances of collar rot due to *Rosellinia* sp. becoming severe.

c. Time of Sowing

Pathogens are able to attack susceptible plants only under certain favourable environmental conditions, For example, downy mildews require continuous high water content for infection. If by some means, the crop is grown in such. a way that it is not attacked by pathogens during its most susceptible period, disease incidence can be considerably reduced. *Rhizoctonia* root rot of gram is severe if gram is sown immediately after the rains. This is due to the fact that the pathogen develops rapidly under higher temperature and moisture conditions, A. slight delay in sowing is helpful in this case.

d. Disease Escaping Varieties

Many varieties of crops prevent themselves from damage because of their growth characters. These varieties escape the attack of the pathogen and do not resist the attack due to their inherent characteristics, for example, early maturing varieties of wheat or pea escape damage due to *Puccinia graminis trifid* and *Erysiplze polygoni* respectively.

e. Selection of Seed and Planting Stock

In India seed potatoes are generally imported in the southern states from the Simla hills for better control of viral and bacterial diseases as in simla hills disease resistant potatoes are researched and produced for all Inida. In the USA, seed growing areas have been shifted to dry pacific regions for crops such as cabbage, turnip, beans and peas for obtaining disease-free seed and controlling such diseases as black leg and black rot of cabbage and turnip. Since many plants grow by vegetative parts, the selection of disease-free planting material forms a very important control measure. The sowing of disease-free seeds in a disease-free field is often a very important control measure for certain diseases, such as red rot of sugar cane, black scurf of potato, or various viral diseases of potato. Tuber indexing has been resorted to for obtaining virus-free seed and protecting crops against the attack of various viral diseases of potato.

f. *Modification of Old Cultural Practices*

The distance between plants in the field time and amount of irrigation, quantity and quality of fertilizers or organic manures, time and method of planting, mixed cropping method, depth of sowing, etc. are some of the cultural practices which effects the seriousness and occurrence of disease of certain diseases. The changes needed in these practices can help to reduce disease incidence and loss of the crop.

Tillage provides a temporary improvement in aeration, accelerates drying and redistributes components in the soil. These environmental and physical changes can predispose pathogens to antagonistic effects of the microbiota. Early and deep ploughing kills cotton roots and favours saprophytes rather than *Phymatotrichum omnivorum*. According to Garren and Duke, if the plant debris is buried deeper by ploughing, powered the pathogen becomes less potent.

Wrong mechanical sowing makes plants prone to infection. This may also result in an unequal and uneven distribution of seed. Overcrowding of seeds and plants may result in damping off of seedlings. The depth of sowing of seeds has an important effect on some pathogens, notably those which attack seedlings. By delaying the emergence of seedlings, deep sowing may help to increase the resistance of a prove crop to a pathogen. Good control of blind seed is obtained by planting infected seed at least 1.25 cm deep and with complete soil coverage.

Mixed cropping is of special significance in Asian countries. This practice helps to effectively check the spread of infectious diseases from one crop to another

Sclerotium rolfsii on pea-nuts is controlled by ploughing under the surface refuse before sowing. This deprives the fungus of a food base since it fungus grow only on or near the soil surface. The effect, therefore, is very much like that of crop rotation. Deep ploughing of soils that contain diseased potato tubers can effectively decrease the amount of inoculum or *Phytophthora infestans*.

Mixed cultivation of cotton and moth *(Phaseolus aconitifolius)* reduces the incidence of root rot of *the* former crop. Mixed crops, such as wheat and barley, wheat and gram, arhar and jowar, cotton and moth reduce the economic losses from diseases.

Management of soil water offers a means to control the associated micro biota-directly, through water potential, and indirectly, through gas exchange of the soil.

Common scab of potato can be managed by keeping the soil water potential near the tubers higher than -0.4 bar during tuber formation . The dry land foot rot of wheat caused by *Fusarium roseum* 'Culm arum' car be controlled by irrigation, partly because the water reduces stress in the host (Papendick and Cook, 1974), and also because the germlings of *Fusarium* lyse or convert into new chlamydospores at water potentials above -10 to -15 bars.

In Israel, the diseases of tomato caused by *Stemphylium botryosum* and *Xanthomonas vesicatoria* were increasd by overhead sprinkling, whereas *Leveillula taurica,* a powdery mildew which thrived in dry conditions, was reduced. In the presence of excess free water, bacterial cells and fungal spores are disseminated easily. Pathogens, such as *pythium* and *Phytophthora* species, are dependent on high soil moisture content. Cereal rusts are more common in heavily irrigated crops or those grow in very wet soil.

2. Exclusion of Inoculum

The main aim of these method is prevent pathogens from infecting virgin area and crop of that area. Seed certification, crop inspection, growing crops in regions unfavourable for pathogen and quarantine measures are some of the measures of preventing the spread of pathogens.

a. *Seed Treatment*

Seeds, tubers, grafts, bulbs and other propagating materials can be given heat, gas or chemical treatment to keep the seeds free of pathogens.

b. *Quarantine Regulation*

A quarantine can be defined as a legal obtainment on the movement of agricultural products for the purpose of exclusion, prevention or delay in the spread of plant pests and diseases in uninfected areas. In most agricultutally advanced countries, plant quarantine rules and regitation has been placed on the statute book to restrict movement of diseased plant material or of fungi, bacteria or viruses that cause diseases in plants. Quarantine measures are of three types:

(a) domestic,

(b) internal, and

(c) total embargoes.

The quarantine laws were first passed in the USA, in 1912 and are known as Federal Quarantine Acts. In India, the Destructive Insect and Pest Act was passed in 1914 and subsequently supplemented by other legislations.

The DIP Act in India was passed in 1914 and has been revised eight times from 1930 to 1956 and corrected up to 1967 but it still is not enough to fulfil the needs under the prevailing circumstances. In India there are 16 quarantine stations operating under the Directorate of . Plant Protection and Quarantine—eight at sea ports, six at airport and two on land frontiers (Hussainiwala in Ferozepur and Sukhiapokri in Darjeeling district). There are a number of gaps in the DIP Act, and as a result many serious plant pathogens have already been introduced in the country. These are now heavily destroying crops in certain areas. Golden nematode and wart diseases of potato have been introduced in India from the European countries and are now well established in the Nilgiri Hills and Darjeeling. Fortunately, timely action has prevented the spread of wart disease to other parts of India.

In India, so far, quarantine acts and regulation exist for two insect pests, i.e. Fluted Scale *(Tcerva purcheri)* and San Jose Scale *(Quadraspidiotus perniciosus)* and for three diseases, i.e. wart *(Synchytrium endobioticum)* of potato, bunchy top (virus) and mosaic (virus) of banana domestic with a view to prevent the spread of these pests.

The problem of plant disease is international. Hence, in 1951 in Rome, an International Plant Protection Convention was drawn up which at present has about 50 signatory nation members. Briefly, each contracting government agrees to make provisions for:

(i) An official protection organization with the specific basis of inspecting, growing crops and the produce derived from them and issuing phytosanitary certificates.

(ii) The distribution of information regarding pests arid diseases both within the country and to FAO so that a world reporting service is established.

(iii) Research and investigation in the field of plant protection on a cooperative basis for disease which had international effects. Within the framework of the international cooperation there are six regional groups.

a. European Plant Protection Organization (EPPO). This was formed prior to the treaty in Rome.

b. Inter African Phytosanitary Commission (for South Africa, South of Sahara).

c. Plant Protection Committee for South East Asia and the Pacific region.

d. Organizmo International Regional de Sanidad Agroperciane (Central America and Mexico).

e. Convenio Intermericano de Protection Agricola (South America).

f. Near East Plant Protection Commission.

3. Eradication of the Pathogen

The main aim of these measures is to break "infection chain" by removing the focus infection and starving the pathogen. A knowledge of the life cycles of the pathogens; method of perennation, host range and habit of growth are essential prerequisites for introducing suitable measures to prevent the disease. Eradication involves totally removing the pathogen from infested areas, and the magnitude of the operation involved may differ considerably. Eradication is attempted through roguing, sanitation, crop rotation, eradication of alternate and collateral hosts, heat and chemical treatment of diseased plants, soil treatments, biological control, etc.

a. *Roguing or total destruction of infected plants*

This practice contains careful removal and destruction of infected plants or plant parts - from the field at an early stage by removing the "foci" of infection and preventing wide spread of the pathogen. Rouging is employed in such diseases as loose smut of wheat, loose and covered smut of barely, sorghum, maize, red rot of sugarcane, ergot of bajra, wilt of arhar, ,and many viral diseases.

b. *Eradication of Alternate or Collateral Hosts*

Many plant pathogens needs two hosts to complete their difference; one of them is generally a wild plant. The destruction of such a wild host plant breaks the chain in the completion of the life cycle of the pathogen. Many plant pathogens have several alternate hosts on which they pass their life cycles. The destruction and removal of collateral hosts is an important control measure of the disease. The yellow vein, mosaic of lady's finger persists on wild hosts *(Hibiscus tetraphyllus)* in nature and the systematic removal of this wild plant helps to control the disease.

c. *Crop Rotation*

Very often, regular cultivation of the same crop or related crops in the same field leads to the perpetuation of a pathogenic soil-borne fungus and the gradual increase in the intensity of the disease. Soils generally suffer from what we call soil sickness. When the same crop is raised year after year in the same field, the soil-borne pathogens of that crop easily establish themselves in the soil and increase their population. After a certain period of time, the soil becomes so heavily populated with the pathogen that it no longer supports the growth of the crop. Such soils are called "sick soils". Crop rotation helps to 'control many soil-borne diseases, such as wheat mosaic, wilt of linseed; finger and toe disease of crucifers, ergot and smut of bajra, bunt of wheat, leaf smut and bunt of rice.

Curl has studied & reviewed the control of plant diseases by crop rotation. This method is also important with other point of view. If the same crop is sown in the same area year after year, there are chances that a particular essential nutrient in the soil will be exhausted. There can also be the accumulation of toxic substances and organic acids if a specific crop is continually cultivated in the same area. Crop rotation is generally recommended which require totally different nutrients and affected by different pathogen.

d. *Sanitation*

Like plant sanitation, field sanitation is also essential for the control of many plant diseases. Pathogens survive through dormant structures in plant organs in the soil. The collection and destruction of these structures is an important control measure. It is common practice to bum diseased stubble for plant disease control. The effectiveness of fire for controlling plant disease has been reviewed by Hardison. In fact, sanitary conditions are helpful in almost all infectious diseases. The spread of certain diseases, for example, tomato mosaic virus, would not be rapid if the workers who smoke were to wash their hands before handling the plants.

e. *Heat or Chemical Treatment of Diseased Plants*

The pathogen present in the plant can be killed or deactivated by applying heat or chemicals. Hot water treatments have generally been most successful in controlling diseases and are used to eliminate *Ustilago nuda* from wheat and barley seed, the nematode *Dictylenchus dipsaci* from *Narcissus* bulbs and, ratoon stunt and other diseases from sugar cane setts.

Treating the seeds with fungicides, especially organomercurials, is new varieties of a common practice to remove the superficially borne inoculum, and there are now systemic fungicides available which can eliminate deep seated fungal infection. More recently, thiram soak has been successfully utilized to eliminate *Septaria appiicola* from celery seed, and *Mycosphaerello pinodes* and *Ascochyta pisi* from pea seed.

f. *Soil Treatments*

Method of soil treatment involves the use of chemicals, heat flooding, etc. Plant pathogens can be killed & removed from the soil by dry heat, steam or hot water, flooding and fumigation with chemicals. Heat and chemical treatments have been successfully used for the eradication of plant pathogens from green house soils as well as soils in plant beds. The treatment of soils in large fields has not been so effective. Nematodes can be killed in the soil by fumigating it with DD mixture or with other nematicide. PCND has been found to be effective in eradicating sclerotia-producing soil pathogens, formaldehyde in controlling onion smut disease.

g. Biological Eradication of Plant Pathogens

The term biological control clearly means control of a disease through some biological agency, and the term 'biological agency', means a living micro-organism or macro-organism other than the diseased or damaged plant acting as host and the pathogen or pest causing the disease or damage. According to Garrett, "Biological control of plant disease may be defined as any condition or method hereby survival or activity of a pathogen is reduced through the agency of any other living organism (except man himself), with the result that there is a reduction in the incidence on he disease caused by the pathogen".

In 1951 a new method was approached by Bliss in the USA who was working on root disease of citrus caused by *Armillaria mellea.* He suggested that if the soil is fumigated with carbon disulphide, the disease producing fungus is killed not directly by the fumigant but indirectly by the produce and dominance of *T. viride,* the population of which gets increased after fumigation. Garrett tested the effect of fumigation and the antibiotic effect of *T viridearid* assessed the relative parts played by these two components. He found that about 30% of *Armillaria mellea* was killed due to the dominance of *T. viride,* while the rest was directly killed by the fumigant. He also pointed out that in the thicker plant roots, *T. viride* may be more effective in the process of killing than the fumigant which might not be able to reach easily.

The most conspicuously successful biological control is

by Rishbeth in England. Rishbetb has been studying a root disease of pines *(Pinus sylvestris* and *P. nigra)* caused by basidiomycetous fungus *Fomes annosus.* The disease spreads through infection by basidiospores produced in the large fructifications. The spores infect the left-over stumps when the trees are felled. Rishbeth at first tried to control. The disease by painting these stumps with fungicidal paints, such as creosote. This brought only partial control because the stump tissue cracked with progressive drying, causing fissures through which the fungal spores of *F. annosus* could enter and cause infections. Rishbeth's research associate, Meredith made an important observation that along with *Fomes annosus* there were a few other fungi which were of the nature of weak parasites that colonized the cut surfaces of stumps. Rishbeth has pointed out that he originally selected *Peniophora gigantea* for this purpose because it is the most common and effective competitor of *F. annosus* in the natural colonization of pine stump surfaces. It decomposes lignin and cellulose vigorously and the fungus establishes itself in the stump quickly and from there into the stump roots. From these stump root infections the pathogen invades the roots of other healthy trees. Rishbet showed that if paste of oidia of *P. gigantea is* pointed on the stump, it capture it very tart and kills the tissues, making it easy for saprophytes to colonize and decompose the stump and making the tissues totally unsuitable for an attack by *Fomes annosus* which is a specialized parasite. Rishbeth determined that a dose of 1 × 10″ oidia of *Peniophora gigantea* allowed a satisfactory margin of safety for effective biological control of the largest spore inoculum of *F. annostls* for a pine stump of 16 cm wood diameter.

h. Soil Amendments

Many root diseases have been controlled by ploughing organic matter into the soil, such Several others have been successfully controlled in this way, such as *Phymatotrichum* root rot of cotton, "take all" *(Ophiobolus graminis),* several *Rhizoctonia* and *Fusarium* diseases.

In the case of *Ophiobolus graminis,* it has been shown that organic amendments, rich in carbon and deficient in nitrogen, control the "take all" disease of wheat. This can be explained as follows: As a result of organic amendment with a high C/N ratio, there is a big increment in the microbial number and activity. There is considerable liberation of CO_2 by soil saprophytes which diminishes the pathogenic activity of this fungus. In the process of survival also, low nitrogen content in the soil reduces the life of the fungus.

The organic amendments increase the activity of saprophytic organisms in a flush which results in the abundant production of CO_2 as a result of the respiratory activity of the micro-organisms. The pathogens sensitive to CO_2 are killed. Due to the rapid multiplication of micro-organisms in the soil, the available nitrogen in the soil is rapidly used by the fast growing saprophytes. This results in acute nitrogen shortage which adversely affects the growth of the pathogens. In their studies on the control of bean root rot with barley amendments, Snyder established that additions of nitrogen nullified control. The germination of the chlamydospores is poor and the hyphae produced are also depauperate. This is perhaps due to the locking up of avail able soil nitrogen in the cells of micro-organisms which decompose the organic material. This is thought to deprive the chlamydospores of the pathogen of even the minimal amount to soluble nitrogen required for germination and host, infection. The pathogen and the disease develop rapidly, if nitrogen is increased. It has been shown that the inability of the conidia to germinate & grow in aqueous extracts of soil solution with cellulose (C/N ratio higher than 2:1) and barley (C/N ratio 55) can be overcome by as small a quantity of nitrogen as 5 ppm. All these factors show that nitrogen limits pathogenesis. Extensive data about the behaviour of *Rhizoctonia solani* have been provided by Papavizas and Davey. Davey and Papavizas for the control of *Rhizoctonia* disease of snap bean *(Phaseolvs vulgaris)* by the use of organic amendments. Blair in studies with Rossy Chlodny slides demonstrated that the growth of *Rhizoctonia solalli* in unsterilized soil is checked by the incorporation of 1% finely divided organic matter, such as dried grass, alfalfa or wheat straw; Concerning the possible mechanism of this biological control, Davey and Papavizas have produced data that largely confirm the explanation originally given by Blair that the activity of *Rhizoctonia solani* is temporarily checked

(a) by an increase in the carbon dioxide content of the soil and

(b) by a scarcity of available nitrogen in the soil solution. Both effects are produced by rapid decomposition of nitrogen, deficient organic matter in the soil. Davey and Papavizas showed that maximum inhibition of competitive saprophytic colonization of buck wheat stem segments by *Rhizoctonia solani* occurred in soils receiving amendments adjusted to C/N ratios 100-40. A lower degree of inhibition was provided by amendments with either higher (400-200) or lower (20. 10 and 5) C/N ratios. This result can be explained by postulating that nitrogen exerts two distinct and opposite effects upon the activity of R. *solani,* the first being predominant down to C/N ratio 40 and the second coming into action at C/N ratios of 20 and below.

The first or inhibitory effect can be attributed, to the action of nitrogen in enhancing the rate of decomposition of the organic material and hence in promoting a more rapid production of carbon dioxide. This increase in. carbon dioxide output can be expected to closely fellow the trend in total microbial numbers from C/N ratios 400 down to 5. At C/N ratio 20, however, the second or stimulatory effect (with respect to *Rhizoctonia solani)* of nitrogen comes into force and the total decreasing effect of the organic amendment upon the fungus is suddenly decreased rapidly. This can be ascribed to the presence of the available nitrogen that can be directly absorbed by the mycelium of *Rhizoctonia solani,* thus removing the ,disability of nitrogen starvation suffered by the fungus in the presence of organic amendment adjusted to higher C/N ratios. This change is exactly where we might have predicted it to have occurred, that is, in the region of C/N ratio of 20. The C/N ratios of fallow soils and of microbial substance is generally around 20.

Mitchell reported that soil amendment with either chitin or laminarin at 1020 kg/hectare reduces incidence of bean root rot and also of radish wilt due to *Fusarium oxysporum* f. *conglutinans.*

Phytophthora root rot of avocado was controlled by Zentmyer by amending the soils with alfalfa meal-a material of low C/N ratio. Similarly, Mehrotra and Tiwari could successfully controlled the root *rot* of *Piper betle* by organic amendments and antagonists growing on organic amendments in the, soil amended with corn straw gave best control. Corn straw supplemented with nitrogen (NH_4NO_3) gave still better control. The effect of different antagonists growing on corn straw and til *(Sesamum indicum)* oil cake amendments showed that *Trichoderma viride* gave best control. Mehrotra and Claudius tried to eradicate the root rot and wilt diseases of *Lens culinaris* by organic amendments of the soil and also introducing antagonists in the soil along with the organic amendments. The wilt of lentil caused by *Fusarium oxysporum* f. *lentis* was checked with change of wheat straw, wheat straw plus ammonium nitrate, corn straw and sugar cane straw. The successful antagonists were *T. viride* and *Streptomyces gougeroti.*

Biological control in potato diseases of underground parts has been studied by Shrivastava and Saksena at Sagar. In soil amendments, the *Rhizoctonia* disease was controlled by nitrogen-rich amendments, such as the tops of bean plants. Carbon rich amendments, such as cellulose and sugarcane straw did not control the disease at all. This can be explained on the basis of increased activity of *Rhizoctonia solani* which readily breaks down cellulose. The wilt disease due to increased *Fusarium solani* could not be controlled in these experiments by organic amendments.

One good examples of biological control with green manuring is that of potato scab with a soybean cover crop in southern California. *B. subtilis* grew equally well in pots of soil amended with either green soybean or green barley tissue, although the bacterium multiplies in much greater amounts on soybean than on barley.

There are many examples where attempts have been made to infest the seeds or propagules with organisms with a view to reduce root disease. Some workers have however emphasized the non content effects of seed inoculation with respect to biological control. Chang and Kommedahl were successful in biologically controlling seedling blight of corn by coating seeds with *Bacillus .subtilis or Chaetomium globosum.*

i. *Hyperparasitism*

Certain fungi and bacteria act as parasites on plant pathogens. The genus *Cinccinobolus* includes species of fungi which attack members of the family, Erysiphaceae. *Cinccinobous* seldom reduces the attacks of the powdery mildews in nature because it very rarely occurs in large quantities. *Cinccinobous* can be grown on artificial media, however, and some good results have been obtained by artificial inoculation. Several workers have tried to utilize *Eudarluca australis* for controlling rusts. The fungus does control the disease under certain natural climatic conditions over which man has no control. As discussed by Chester extra inoculum of the hyperparasites will not be of much use as the weather, unless very favourable *to* the development of this hyperparasite, is unlikely to permit any marked control' of the disease. There are several other fungal species which attack rusts, such as *Tubercularia* spp. which attacks aecia and spermogonia of several rusts and *Verricillium hemieiae* on *Hemileia vastatrix.*

Many fungi are known to parasitize phytopathogenic nematodes. In 1934 Rozsypal and Schmidt reported a *Protomycopsis chytridiale* which attack *Heterodera schachtii.* In 1933, H. Goffart recognized that the eggs and embryoes of *Heterodera avenae* could be parasitized by an ascomycete, *Cylindrocarpon radidcola,* which is known to attack the underground parts of various cultivated plants.

Hyphomycetes as nematode predators are numerous, and are frequently met with in nature.

Several attempts have been made to use fungi for the control of nematodes, such as Meloidogynes which attack pineapple in Hawai for nematodes parasitic on animals in France and for controlling the golden nematode of the potato in England. Some good results were obtained but practical use of these fungi in agriculture is still a far off possibility.

j. *Bacteriophages*

In 1917 d' Herelle presented proof of a transmissible lytic principle that acted on the I shiga bacillus, showing that bacteria can be diseased too. A number *of* phages have now been discovered for many phytopathogenic bacteria such as *Agrobacterium tume faciens, Erwinia aroidea, E. carotorora, corona. fadens, p, glycinea, P. tahaci. P. xanthoclona, Xanthomonas citri, X. phaseoli, X. solanacearum. X. oryzae,* etc. but not much success has been obtained *to* control these diseases at big scale with the help of their respective phages.

k. *Entomogenous Fungi*

There are many fungi which are capable of growing in the bodies of insects and thus destroying them. Scientists are looking for ways of protecting crops against insect pests. With the introduction of DDT, it was felt that a way had been found to destroy insects affecting crops. DDT and scores of other insecticides that, were introduced killed not only all the insects but also the predators, including fish in water (the insect enemies), with the result that in some cases more damage was done more than any benefit has occurred. Plant pathologists can use micro-organisms which are immune or at least not affected by pesticides for biologically controlling insects. One *of* the important fungus which has been utilized is *Entomophthora.* About 100 species are known, mostly as insect parasites. While some appear to be rather specific in their host range, others are capable of-attacking a wide range of hosts. Some species have been grown in culture.

USE OF CHEMICAL METHODS IN DISEASE CONTROL

Plant diseases play a big role in costing of food. Plant pathology must solve the food problem by devising new method measures and improving the existing ones. The annual loss to world crops as a result of disease has been estimated at

25,000 million dollars; of this, a major part is due to fungal pathogens. Surface deposits of fungicides can control many fungal diseases if properly applied, but it appears that more effective control can be achieved with systemic fungicides which can penetrate and move in the plant.

The use of chemical sprays, dusts or seed treatment for protecting plants from the attacks of the pathogens is not an invention of the 20th century. The early agriculturists try to control plant diseases by utilizing chemicals and fungicides. The first recorded mention of plant disease control is in the writings of the Greek poet, Homer (1000 B.C.), who mentions sulphur, which is still in use. Also, the fumigation of trees with bitumen and sulphur was mentioned by the Roman patriot, Cato.

The first main landmark in the control of fungal diseases of plants was the discovery by Anton de Bary that the causal agents of many plant diseases are fungi. The progress of fungicides quickly followed this discovery. It is about this time that Louis Pasteur discovered the bacterial parasite of silk worm. He proved the real nature of fermentation and put an end for ever to the doctrine of spontaneous generation. It is after this period that Pierre Alexis Millardet showed that the downy mildew of grapes could be controlled by mixtures of copper sulphate and lime. It was only ip. 1885 that its toxicity was reduced by mixing it with lime. The Bordeaux mixture was used both in Europe and the USA for the control of many diseases. Substances used as late as the middle of the 19th century had, offensive or caustic properties, acid or bitter taste and pungent odour. These were regarded as good qualities in a fungicide. The Bordeaux mixture was followed by lime sulphurs in the late 19th century, and formalin, copper carbonate dusts and organomercurials as seed treatments. This gave way to a period of organic fungicides-thiram, chloranil, dichlone were developed from the mid-1930s onwards, and captan, one of the most successful ones, was developed only two decades ago. The importance has shifted to the oxides of copper. Many groups of chemical compounds are now available with the plant pathologists for plant disease control.

These include heterocyclic nitrogen compounds, quinones, phenols and antibiotics. There is increasing interest in the chemotherapy of plants, and systemic fungicides are of .current interest to Scientist.

Before discuss the various chemicals for plant disease control it is important to define a fungicide. "A fungicide is an agent that kills or inhibits the development of the fungus spore or mycelium". The fungicides used of plants may be classified as protectants, eradicants and therapeutants the basis of their uptake by and mobility within plant tissues.

1. Protectants

As the name suggests, protectant fungicides are prophylactic in their nature. Protectant fungicide, which may be applied to seeds, plant surfaces or the so cannot penetrate plant' tissues in effective amounts. Therefore, they act outside the plant prior to infection, by the pathogen.

2. Eradicants

These chemicals kills the dormant or active pathogen from the host. They can remain effective on or in the host for some time and function as protectants also. Chemicals can be protectant as well as eradicant.

3. Therapeutants

A therapeutant is an after agent that inhibits or stops the progress of a Disease syndrome in a plant when applied after invasion by a pathogen. Therapy can be by physical means, such as solar energy treatment or hot water treatment of smutted wheat grains, but generally it is by chemical means and is then called chemotherapy. Usually the chemotherapeutants are systemic in their action, that is, they enter the plants and affect deep-seated infection.

To be successful, fungicidal sprays must have some constant features. To be effective against rain and dew they must be nearly insoluble in water, but at the same time, they must be soluble enough to function as fungicide. Tenacity or adhesiveness to host plant is another quality needed to enable them to stick well and last for a longer period. Further they should not be poisonous to the host parts (phytotoxic) so that they can be used safely. To meet all these requirements, various components are mixed with the fungicides; which are known as stickers spreaders and safners.

The fungicides are can be classified following categories:

a. Inorganic copper compounds

b. Inorganic mercury compounds

c. Sulphur sprayes

d. Organic compounds

e. Quinone and Phenolic fungicides

f. Heterocyclic nitrogen compounds

g. Benzene compounds

h. Organomercurials

i. Systemic fungicides

j. Antibiotics

k. Fumigants

l. Oils

Inorganic Copper Compounds

(i) *Bordeaux Mixture*

In 1882, Millardet, studying the downy mildew of the vine a disease then recently introduced in France from the USA, saw that along the roadside at Medoc in Gironde, certain vines had leaves while others had been defoliated by the disease. Millardet believed that the foliage on some of the vines had persisted because of the dabbing of the leaves with verdigris or with a mixture of lime and copper sulphate. Following this hint, Millardet in collaboration with Gayon, a chemist, was able to announce in 1885 the successful use of a lime copper sulphate mixture as a fungicide against *Plasmopara*. This mixture has the reaction product of copper sulphate and calcium hydroxide. This discovery has doubtless been the greatest boon to plant protection since about fifty million kg of copper sulphate is commonly used in the USA for this purpose every year.

The formulae for the Bordeaux mixture is different for different chemists. The original mixture had 15% copper sulphate and 8% lime (CaO). The percentage of these now varies according to the crop and the disease. Higher concentrations are useful in the case of woody and tough tissues, while lower concentrations are used for succulent and softer parts. The following formula (5:5:50) is regarded as a good standard type: It is also called Bordeaux mixture.

Copper sulphate (blue stone)	5 lb (2.268 kg)
Stone or hydrated lime	5 lb (2.268 kg)
Water	50gallon

The mixture is prepared by mixing $CuSO_4$ solution with lime stone solution. The methods of preparation are given by Snarvelle. The two solutions are prepared separately in wooden or earthen vessels and the diluted solution of $CuSO_4$ is slowly poured into a concentrated lime solution in a tub while stirring continuously. The resulting mixture is neutral or slightly alkaline and is not injurious. The mixture is used almost immediately after preparation and deteriorates if left for some time. To avoid this, stock solution of the two ingredients can be prepared separately and stored in wooden vessels. These can be mixed when actually required.

The chemistry of Bordeaux mixture is complex and is not fully understood. The expected reaction is:

$$CuSO_4 + Ca\,(OH)_2 = Cu\,(OH)_2 + CaSO_4$$

It is not cupric hydroxideas would be expected by a straight forward reaction of one molecule of copper sulphate with one molecule of lime. The blue hydrogel contains tribasic copper sulphate [$CU_4\,SO\,(OH)_6$] with absorbed lime and free lime and calcium sulphate. With absorbed lime, substances of various compositions are formed, depending upon the ratio of lime to copper sulphate, their total concentration and the period of time that has been spent after mixing. Eventually the blue hydrogel contains a series of compounds in the process of transition rather than one final product.

The cupric hydroxide is the active component which is soluble and is toxic to spores and sporelings. The Bordeaux mixture is a suspension which is fairly tenacious, nearly insoluble and spreads quite well. It may be detrimental or beneficial to different hosts, Beneficial effects have been observed on grape, potato and other crops. Greener foliage is often the outstanding result and it also has insect repellent properties.

The Bordeaux mixture is sometimes toxic to tender leaves and other parts and bums them as well as fruits and other aerial parts. Sometimes spots occur, as on apple fruit. The fungicidal action is due to the formation of soluble copper, which is toxic to spores or sporelings in drops of atmospheric water. The factor which brings about the solubility of toxic ions is still not fully understood. There are three possibilities:

a. Carbon dioxide and ammonia in meteoric water.

b. Secretions by the fungus either before or after spore germination

c. Secretion from the host plant .

Or all these three processes might be instrumental in the solubility of copper. McCallan has reviewed the literature on the fungicidal action of copper.

The Bordeaux mixture is specific against downy mildews, such as late blight of potatoes, koleroga of areca nuts, coffee rust, various leaf spots diseases, blights, anthracnoses, etc. It has remained the standard fungicide against downy mildews of grapes and other hosts since its inceptio in 1882.

(ii) *Burgundy Mixture*

Burgundy mixture is a modification of the Bordeaux mixture, it is often used in cases of tender foliages, since it is less phytotoxic. In this, lime is substituted by sodium carbonate, and hence, is sometimes called 'soda bordeaux'.

Copper sulphate	101b (4.536 kg)
Sodium carbonate	12.21b (5.666 kg)
Water	50 gallons

Burgundy mixture does not produce blemishes or spots. It should be mentioned that neither burgundy mixture nor other substitutes of Bordeaux mixture have been widely accepted.

(iii) *Cheshunt Compound*

This compound was suggested by Bewley in 1921. It is made up of two parts of copper sulphate I and 11 parts of ammonium carbonate. The two substances are well powdered and thoroughly mixed and the dry mixture is stored in an airtight container for 24 hours before being used. If it is left in an exposed condition it gradually loses ammonia and becomes less effective. When spray solution is to be prepared, one ounce of the mixture is dissolved in a little hot water and two gallons of the solution is made by adding cold water. Cheshunt compound is recommended for the control of 'damping off' diseases as a soil drench.

(iv) *Chaubattia Paste*

This paste was developed by Singh at the Government Fruit Research Station, Chaubattia in the Almora district of Uttar Pradesh in India. The mixture is prepared by mixing copper carbonate (800 gm) and red lead (800 gm) in one litre of lanolin or raw linseed oil. This mixture was developed as a wound dressing fungicide to be applied to pruned parts of pears, apples and peaches' for the control of diseases, such as. Stem-black *(Coniothecium chomatosporum),* Stem-brown *(Botryosphaeria ribis),* of apples and pears, and collar rot of apples, peaches, apricots and plums *(Rosellinia* sp.).

Inorganic Mercury Compounds

Many mercury compounds are highly effective as fungicides and bactericides. The use of inorganic mercurial compounds is limited because of high toxicity to human beings. Mercuric chloride ($Hg\ Cl_2$) and mercurous chloride

($Hg_2 Cl_2$) are used as 1:1000 dilutions for soaking the seed, rhizomes and corms of vegetables and flowers to mainly control certain bacterial and fungal diseases. It is also used as a preservative of wood and herbarium specimens. Mercurous chloride is fairly insoluble in water and can be used in the treatment of brassica seedlings against the pathogen responsible for its club root disease. Inorganic mercurial fungicides have been replaced by organomercurials. Inorganic fungicides have been discussed by Martin.

Sulphur Sprays

Its effective use was discovered only in the beginning of the 19th century. Sulphur is the earliest known fungicide and is used widely. Various formulations of elemental sulphur or crude lime sulphur preparations were developed in Europe and the *USA* during the early part of the 19th century to control powdery mildews on a variety of crops and leaf curl of peach. Forsyth, in 1802, appreciated that sulphur was an effective fungicide and recommended its use for the healthy management of fruit trees. It is best known for its effectiveness against powdery mildews on many plants, but is also effective against certain rusts, leaf blights and fruit rots.

(i) *Lime Sulphur*

Lime and sulphur; a mixture which was used on apple against scab, has an added advantage in that it is an insecticide also. Chemically it is a mixture of calcium thiosulphate and calcium polysulphides, which are the chief components. A major part of the polysulphide is converted into free sulphur after lime sulphur is sprayed on the foliage. This free sulphur is the active principle. The polysulphides are thought to be phytotoxic

Lime sulphur is toxic, especially on the tender parts, causing foliage and fruit burns and hence it has only limited use. It has been replaced by self-boiled lime sulphur which is less toxic.

(ii) *Self-boiled Lime Sulphur*

This is made by using 8 lb of sulphur flour and 8 lb of burnt lime with 50 gallons of water. A small amount of water is added to the lime and when well shaken, sulphur is added and the ingradients are mixed vigorously. After boiling has ceased, the remainder of the water is added. This material is less toxic than the lime sulphur described above.

Even this fungicide is now sparingly used and is being replacccd by wettable table sulphurs.

Lime sulphur has some eradicant properties also and is active against apple scab and powdery mildew.

(iii) *Wettable Sulphurs*

These are very finely divided (micronized) elemental sulphurs, also called floatation sulphurs, which are made from hydrogen sulphide gas from the coke industry. These have come into large use because they cause less injury to leaves. Wettable sulphurs are used against apple scabs, powdery mildews of grape and other crops, ornamentals ,and vegetables. In green houses it is used against rusts also.

The mode of action of sulphur as a fungicide is still not clear. Originally it was assumed that sulphur acted directly on the protoplasm of the fungus spore or mycelium and destroyed it. Later it was suggested that sulphur acted in the form of H_2S. Still others thought that it acted as sulphur dioxide. The most recent evidence suggests that sulphur acts as a hydrogen acceptor and thus interferes with the normal hydrogenation and dehydrogenation reaction of the cell.

Organic Compounds

During the second world war there was a shortage of metallic compounds of Cu and Hg which gave an importance to the development of organic compounds. Tisdale and Flenner (1942) first demonstrated the fungicidal possibilities of the carbamates in 1931 in the laboratories of E.I. Du Pont Company, USA but production for vast scale maridel use started a decade later. It was in 1934 that the patenting of dithiocarbamates was announced by Tisdale and Williams by patenting 'thiram' on the use of this group of materials as insecticides and fungicides. Since then much work has been done on the chemistry of thiocarbamates.

Carbamic acid is derived from carbonic acid by the replacement of a hydroxyl group by an amino group.

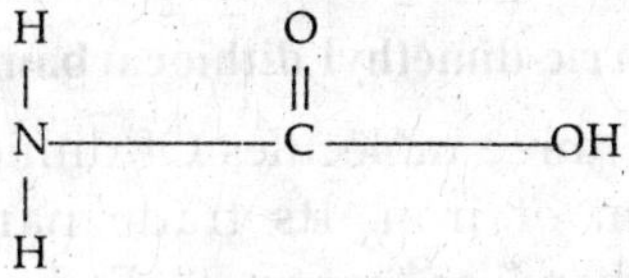

```
H      O
|      ||
N------C--------OH
|
H
```

Carbamic acid

Dithiocarbamic acid is derived by the replacement of two oxygen atoms by sulphur, giving the hypothetical dithiocarbamic acid.

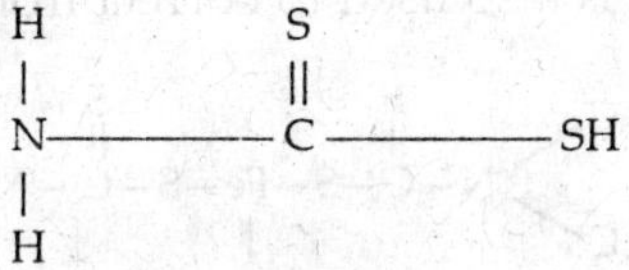

```
H      S
|      ||
N------C--------SH
|
H
```

Carbamic acid

From dithiocarbamic acid the following important fungicides have been derived.

(i) *Thiram* (Tetrametltyl thiuram disulphides) CTMT D3

TMTD is a condensation product of dithiocarbamic acid. This was used as an activator in the production of synthetic rubber.

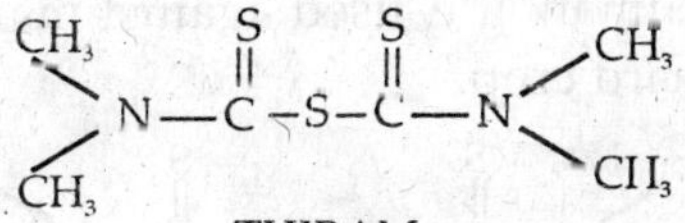

```
CH3       S     S        CH3
   \      ||    ||      /
    N---C--S--C---N
   /                    \
CH3                      CH3
```

THIRAM

Thiram is being sold under a variety of other trade names, such as 'arasan', 'tundas', 'terson' etc.

It is a big seed treatment for wheat, rice, gram, peas, mustard and onion smut. It is also used in the control of foliage diseases and acts as an insecticide. Among the soil-borne pathogen controlled by it are *Pythium, Rhizoctonia, Fusarium* and *Protomyces.*

As a foliage spray, thiram was not as successful as its metallic compounds. For this reason, several compounds have been developed which are as follows:

(ii) *Ferbam* (Ferric dimethyl dithiocarbamate)

In ferbam, three molecules Of dithiocarbamic acid react with one atom of iron. Its trade name is "fermate" or "ferbam", "karbam" or "çoromet". Ferbani is compatible with most pesticides, but its use with copper ,lime or mercury compounds tends to reduce fungicidal efficient Ferbam has been successfully used as a protectant fungicide, against a wide variety of fungus diseases of fruits and vegetables. As a foliage spray it is also used to correct iron deficiency.

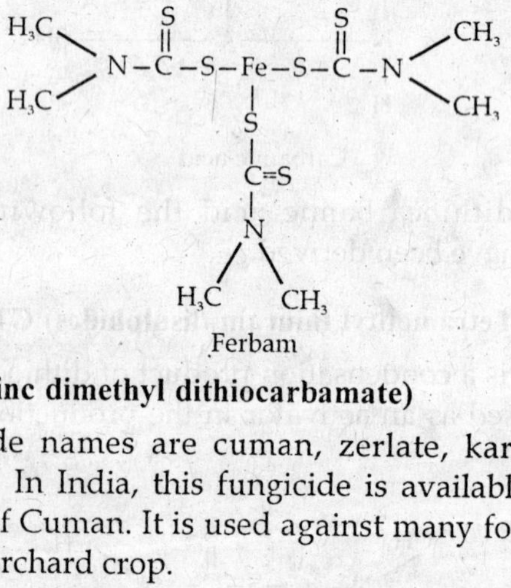

$$(H_3C)_2N-C(=S)-S-Fe(-S-C(=S)-N(CH_3)_2)-S-C(=S)-N(CH_3)_2$$

Ferbam

(iii) *Ziram* (Zinc dimethyl dithiocarbamate)

The trade names are cuman, zerlate, karbam white, corozate, etc. In India, this fungicide is available under the trade name of Cuman. It is used against many foliar diseases in field and orchard crop.

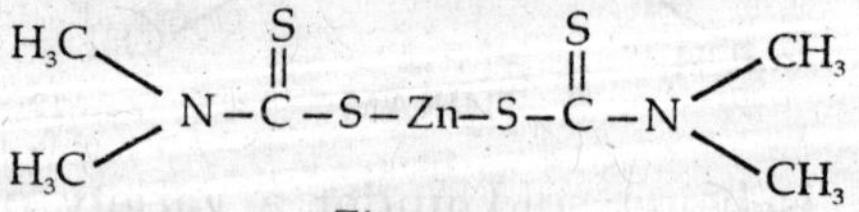

$$(H_3C)_2N-C(=S)-S-Zn-S-C(=S)-N(CH_3)_2$$

Ziram

(iv) *Nabam* (Oisodium ethylene bisdithiocarbamate)

This is sold as Oithane D-14, Parzate liquid or "Dithane A-40". Nabam effective against many leaf blights, such as

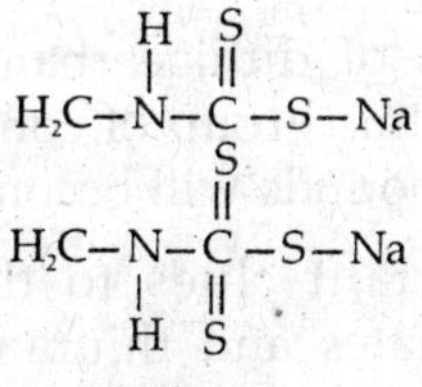

```
      H   S
      |   ||
H2C—N—C—S—Na
          S
          ||
H2C—N—C—S—Na
      |   ||
      H   S
```

Nabam

early and late blight of potato, and seedling blights caused by *Pythium* and *Fusarium* species.

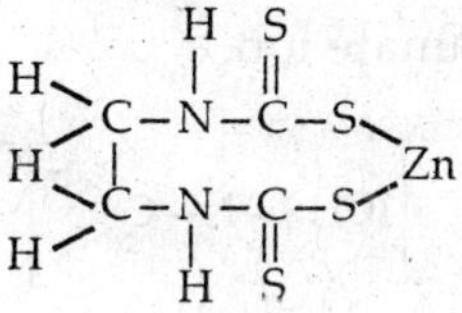

```
          H   S
          |   ||
H \      |   ||
   >C—N—C—S\
H /  |          Zn
   \C—N—C—S/
H /      |   ||
          H   S
```

Zineb

(vi) *Maneb* (Mangnese ethylene bisdithiocarbamate)

(a) This is called as Dithane M-22, Manzate or Dithane M-45. Maneb is best for the control of foliage and fruit diseases *of* many vegetables, flowers, trees; turf and some fruits. It is especially used in vegetable crops as a spray material. (b) Some of the diseases controlled by zineb are bean anthracnose caused by *Collectotrichum lindemuthianum*, downy mildew and

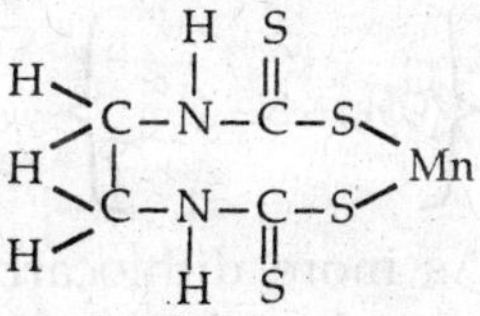

```
          H   S
          |   ||
H \      |   ||
   >C—N—C—S\
H /  |          Mn
   \C—N—C—S/
H /      |   ||
          H   S
```

Maneb

anthracnose of cucurbits, fruit rot of chillies, citrus greasy spot caused by *Cercospora* spp., lettuce downy mildew, maize leaf blight.

(c) In all these dithiocarbamates, the usual doze is 1 kg in 500 litres of water.

(d) These carbamate compounds are very widely used both as dust and sprays for a very wide variety of crop diseases, such as *Altema ria* blights of potato and tomato, *Colletotrichum* anthracnose of tomato, cucurbit, bean, etc.

(e) The success of dithiocarbamates as fungicides has started research on this group of compounds with the result that still better compounds will become available.

(f) Two important clues to the toxic action of the dialkyldithiocarbamates and thiuram disulphides are their formation of metal chelates and their polymodal dosage response.

The zone of inversion where increasing dosage actually depresses toxicity has been not clear in studies. The dimethyldithiocarbamate ion.

$$\left[(CH_3)-N-C\begin{matrix}\diagup\!\!\!\!\diagup S\\ \diagdown S\end{matrix}\right]^{-}$$

react with small traces of copper in the medium to give the highly toxic 1:1 or half chelate,

$$\left[(CH_3)-N-C\begin{matrix}\diagup\!\!\!\!\diagup S\\ \diagdown S\end{matrix}\right]Cu^{+}$$

This is the toxic species. But as the amount of copper in the medium is constant, the addition of more dithiocarbamate forms the less toxic 1:2 or full chelate,

$$\left\{(CH_3)-N-C\begin{matrix}\diagup\!\!\!\!\diagup S\\ \diagdown S\end{matrix}\right\}2\,Cu$$

Toxicity falls. As more dithiocarbamate is added, the intrinsic toxicity of the dimethyl dithiocarbamyl ion exerts itself, acting perhaps by adding essential metallic micronutrients. Thus, toxicity increases again.

The toxicity of ethylene bisdithiocarbamates is believed to be due to the production of isothiocynate which reacts with SH groups.

Quinone and Phenolic Fungicides

Though not as important as the carbamates, quinone and phenolic compounds are quite important as fungicides in certain aspects.

(i) *Chloranil* (Tetrachloro-p-benzoquinoe)

It was reported by Cunningham and Sharvell (in 1940) to give protection to lima bean seedlings from damping off. Chloranil is mainly used in seed treatment. It is sold as spergon and is used in seed and bulb treatment of, legume flowers and vegetables. It is of limited value as foliage fungicide partly because it decomposes in light. It is also used as a soil drench.

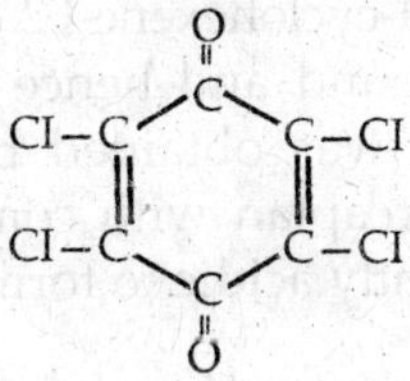

Chloranil

(ii) *Dichlone* (2,3 dichloro-1,4 naphthaquinone)

It was developed later on. Though it is used for seed treatment, it is also useful as a foliage spray against many plant diseases and is apt to cause dermatitis.

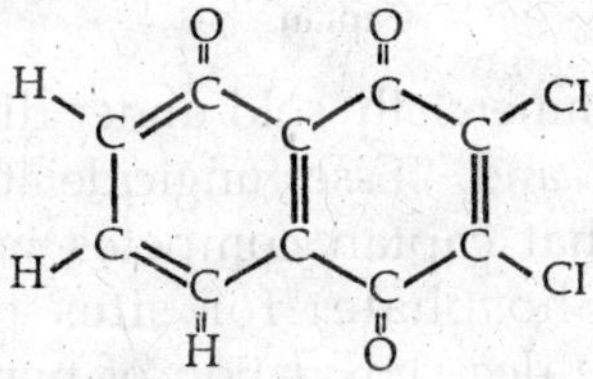

Dichlone

It has been reported that this is four to eight times more effective as chloranil the protection of legume and cotton seed.

(III) *Phenols*

Phenols have shown very good qualities as bactericides, but as fungicides their use has been limited. Phenols have been chiefly applied to textiles and woods as preservative fungicides', but have also been used as soil fungicides and issued treatments. Derivatives of cresol and O-phenyl phenol have been mainly used as preservatives. The bisphenols are a particularly interesting group of phenolic fungicides.

Heterocyclic Nitrogen Compounds

These compounds have a number of valuable fungicidal properties. Their principal uses are as foliage protectants and eradicants of fruits and vegetables.

(i) *Captan*

Captan is the common name adopted for N-trichloromethyl thio-4-cyclohexene-l,2 dicarboximide. Kittleson prepared this compound and hence it is sometimes called, Kittleson's Killer. It was obtained by the condensation of perchloromethyl mercaptan with compounds possessing an imido group sufficiently acidic to form a stable sodium salt.

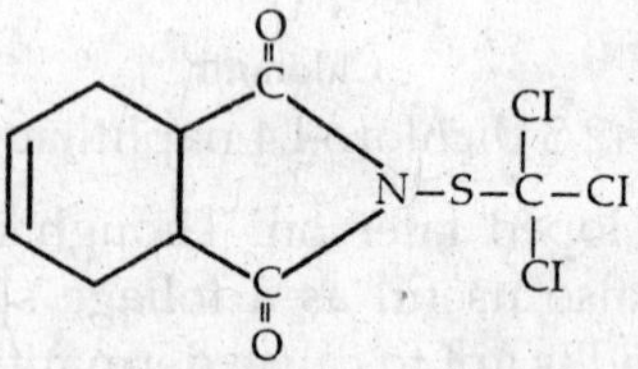

Captan

Captan is commercially sold under different names, such as "Captan 50 W" and "Esso fungicide 406", Hochstein and Cox suggested that captan competes with co-carboxylase (thiamine pyrophosphate) for sites on coenzyme-free carboxylase in the decarboxylation of pyruvate, and thereby interferes with the process of decarboxylation. Another theory of captan action' was presented by Lukens and Sisler. They found that captan interacts with - SH compounds, such as cystein and glutathione, to give thiophosgene. Although a fairly broad spectrum fungicide, it is not generally effective against rusts, powdery mildews and downy mildews. Owens and Blaak showed that captan inhibits many oxidative enzymes and considered that its basic toxic action was an interaction of the intact captan molecule with free sulphydryl groups.

(ii) *Difolatan* [N-(l, 1 ,2,2-tetrachloro ethyl sulfenyl)-cis 4-cyclohexene. 1,2-dicarboximide] It is sold as difolatan and has properties similar to captan. It is a good fungicide for the control of early and late blight of potatoes and tomatoes.

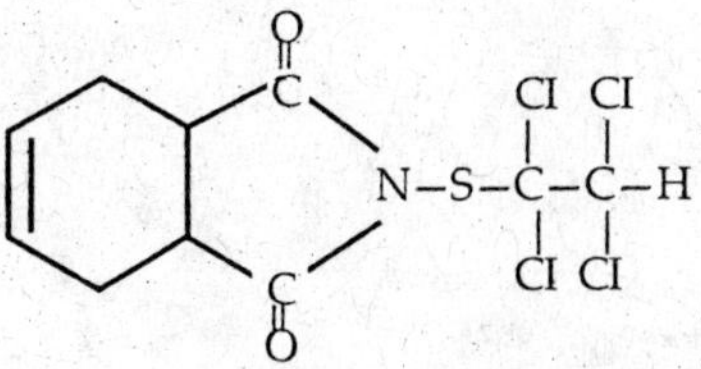

Difolatan

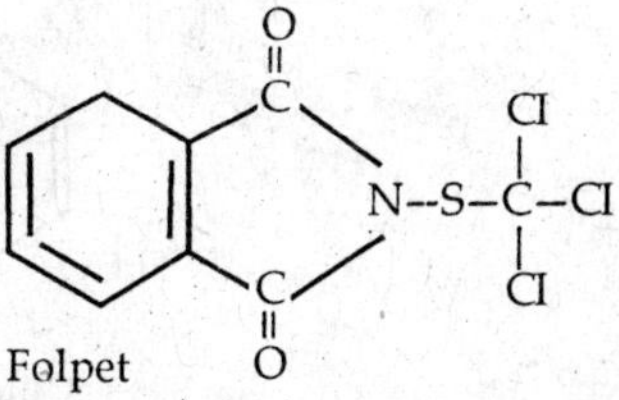

Folpet

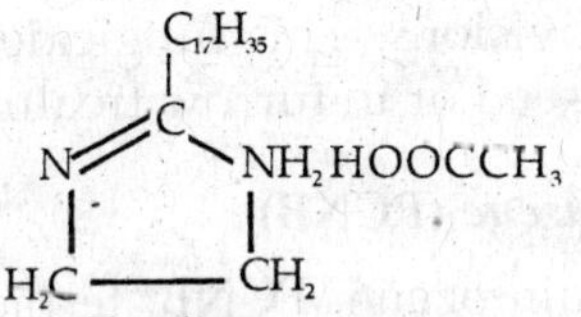

Glyodin

(iii) *Folpet* (N-trichloJomethylthio phthalimide)

Folpet is effective against *Sphaerotheca pannosa.* It is sold under the trade names of Phaltan, Orthophaltan, etc.

Benzene Compounds

Many benzene compounds are poisonous to micro-organisms and several have been developed as fungicide. These have been discussed by Corden (1969). Dinitro-O-cresol is contained in the formulation called Kerenite or Eigetoi. It is used as a dormant spray for the control of many diseases of fruit trees and ornamental plants and for the treatment of wounds in trees.

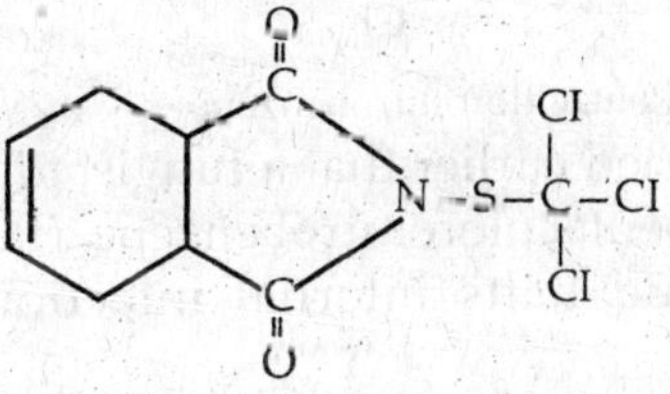

Captan

(i) *Chloroneb*

The fungicide, chloroneb, 1,4-dichloro-2,5-dimethoxy benzene, was introduced by Ryker for the control of seedling diseases of cotton. It has been found active against *R. solani,*

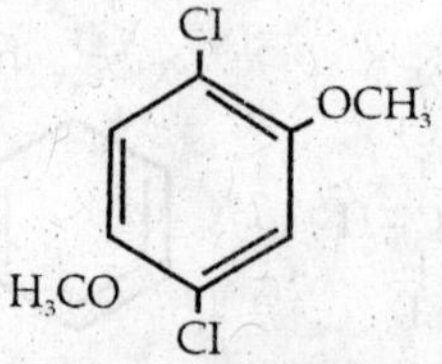

Cl, OCH_3, H_3CO, Cl

Chloroneb

S. *rolfsii* and *Phytophthora cinnamoni.* This is a systemic fungicide and is unusual in that its activity does not follow clear-cut taxonomic divisions. This fungicide has been used for the most part as a seed or in furrow treatment of cotton.

(ii) *Pentachloronitrobenzene* (PCNB)

This is sold as quintozene, PCNB, terrachlor, brassicol, etc. The chlorinated nitro benzenes were introduced as fungicides in 1935 by Brown. They control many soil-borne diseases. PCNB acts by preventing sporulation. It is a vapour phase fungistat which does not prevent spores from germinating but does stop hyphal growth. Horsfall suggested that such compounds act as antimetabolites. It may be that the PCNB acts specifically as a competitive inhibitor of inositol, a required growth substance for many fungi. PCNB also has nematicidal properties.

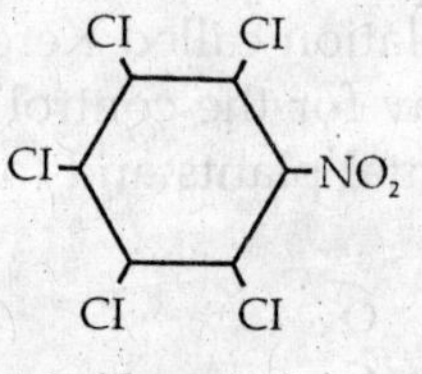

Cl Cl, Cl, NO_2, Cl Cl

Pentachloronitrobenzene

It has been reported earlier that a fungistatic preparation containing 15-20% pentachloronitrobenzene (PCNB) arrests epiphytic growth and halts internal infection caused by *Rigidoporus lignosus.*

Diseases caused by *Rhizoctonia, Sclerotium,* etc. are controlled by PCNB.

(iii) *Dexon*

This is p-(dimethylamino)-benzene-diazosodium sulfonate and is used against soil-borne phycomycetous, fungi

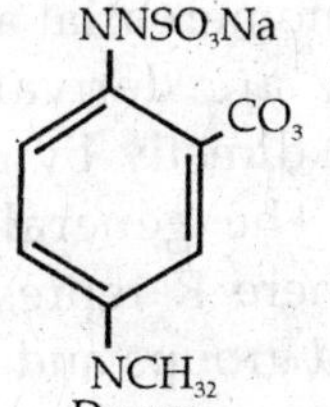

$NNSO_3Na$

CO_3

NCH_{32}

Dexon

causing root rot and damping off diseases of plants.

***(iv) Dinocap* (Karatha)**

Dinocap or Karathane is a mixture of isomers of 2-(I-methyl-heptyl), 4-6dinitrophenyl crotonate. Dinocap is an excellent substitute for sulphur for the control of powdery mildews and is highly specific against them. It is also effective against mites. Actually it was developed as an acaricide and is now extensively used against powdery mildews.

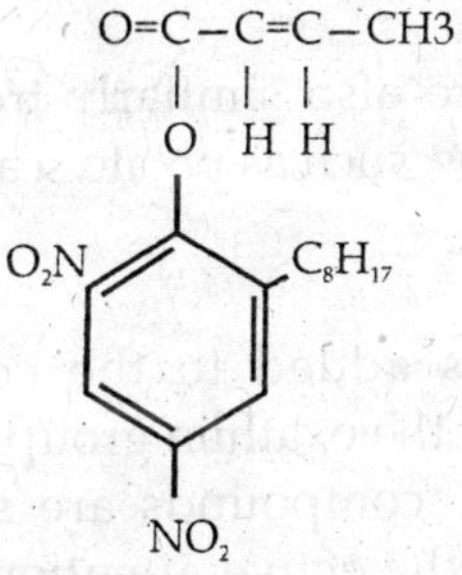

O=C–C=C–CH3

O H H

O_2N C_8H_{17}

NO_2

Dinocap

(v) *Non-aromatic Organic Compounds*

This group includes dodine (n dodecylguanidine acetate $C_{12}H_{25}$ NH. C. (= NH). NH_2 CH_3 COOH) used for the control of apple scab and several other diseases caused chiefly by Ascomycetes and Fungi Imperfecti, as reported by Byrde. Due to its toxicity to some crop plants, its use is limited.

Organomercurials

Organomercurials were developed due to efforts less toxic substitutes for inorganic mercurials. The first important commercial product was 'Uspulum' of the Bayer Company of Germany in 1915. Organomercurial are very widely used for seed treatment. They are derivatives in ,which the mercury atom is attached directly by one or both valency bonds to carbon atoms. The general structure of these derivatives is R. Hg X where R represents a hydrocarbon with or without substituted groups and X represents an acid radical, such as chloride, nitrate, acetate, benzoate, gluconate, etc. R -is usually an ethyl or phenyl group. Some of the important compounds are Ceresan M (ethyl mercury p-toluene sulphnamide), Panogen (methyl mercury dicyandiamide), new improved Ceresan (ethyl mercury phosphate). The chemicals can be used either as dips or dusts. Organomercurial fungicides are mainly used to control externally seed-borne diseases since these compounds are able to eradicate the inoculum from the seed. Their use in the control of externally seed-borne real smuts has been very widely used.

Tubers and bulbs are also similarly treated with dusts against adhering organisms, such as potato scab and black scurf.

Systemic Fungicides

A new chapter was added to the control of fungal pathogens of plants when the oxathiin group of chemicals was produced in 1963. These compounds are systemic in their action and are receiving the active attention of pathologists. Most systemic compounds are absorbed by the roots and transported acropetally to the xylem. They are usually not transported downwards through the phloem tissue.

Major advances in plant disease control will now be made—from systemic fungicides compounds which penetrate and move from their site of application to other parts of the plant. A systemic fungicide could eradicate established infection and protect the new parts of the plant as they are produced, and it would not be subject to loss by weathering. They can be applied in solution as spray on the foliage, but application to foliage is less likely to be effective against the pathogens responsible for root diseases. Several systemic fungicides have been tested as seed dressing to eliminate seed infection. Systemic fungicides have not been very successful in the case of trees and bushes. From the practical point of view, the systemic offer unique possibilities in the tropics, since the active spray residues are not subjected to tropical weathering, thus permitting fewer applications.

Action of Systemic Fungicides

The following are the possibilities of mechanism of systemic fungicides.

a. Inactivation of the enzymes and toxins of the pathogen.

b. Selective accumulation of the fungicide due to greater permeability of the fungus cell wall.

c. Damage to the membranes of the fungal hyphae and inhibition of structures, such as appressoria, cushion formation, emergence of germ tubes and formation of haustoria.

d. Inhibition of fungal enzymes or their destruction. Systemic 'fungicides are more specific in their action than non-systemic fungicides.

Example s of Systemic Fungicides

(i) Benomyl

This is also marketed as benlate. This is methyl-N (butyl carbamoyl) 2 benzimidazole carbamate. Benlate was discovered by E.I. du Pont de Nemours and Co. Inc. in 1968. The benzimidazoleshave been very effective in the control of *Mycosphaerella* on banana and peanut. The systemic activity of benomyl, the most striking fungicide at present, was first reported by Delp and Klopping. Benomyl hydrolyses in acidic solution to methyl benzimidazole carbamate (MBC). MBC has been shown to be the principal fungitoxicant of benomyl which is Systemic in plant tissue (Clemons and Sisler, 1969; Peterson and Edginton, 1970; Biehn and Dimond, 1969).

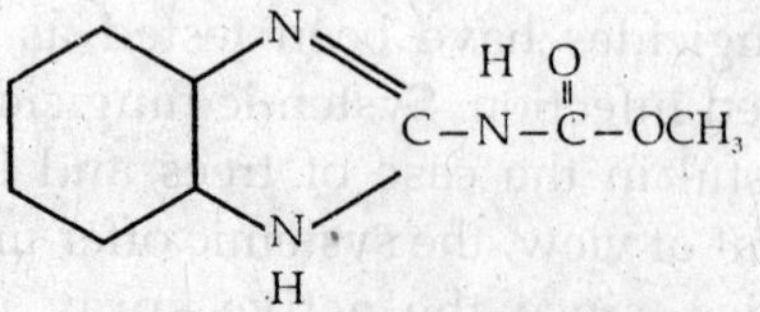

N
H O
C–N–C–OCH$_3$
N
H

Methyl Benzimidazole Carbamate (MBC)

Benomyl has been found effective against a wide range of pathogens. It has a high degree of mite-ovicide activity. It is effective against *Cercospora* leaf spot of sugar beet, rice blast, apple scab, powdery mildew on cucurbits, cereals, and legumes. It is ineffective against phycomycetous pathogens. Many soil-borne pathogens can be controlled, provided the roots are confined to the benomyl treated soil.

Another benzimidazole derivative is thiabendazole, a thiazolyl compound originally developed as an antihelmintic drug by E. Merck and Co. Inc. In 1964, Staron and Allard described its systemic fungitoxicity for the first time. Another derivative is *Baristin.* This is 2-(Methxy-carbamoyl)-benziinidazole which is effective against Ascomycetes and Fungi imperfecti which are considered to be hard to control. It is ineffective against phycomycetous fungi.

Pavistin is translocated within the plant and has both prophylactic and curative actions.

(ii) *Terrazole*

TerrazoJe-5-ethoxy-3-trichloromethyl-l,2,4-thiadiazole is toxic to phycomycetous fungi, such as *pythium* and *Phytophthora.* It has a systemic action in plants (AI-Beldawi and Sinclair, 1969). Little is known about its fungitoxic action.

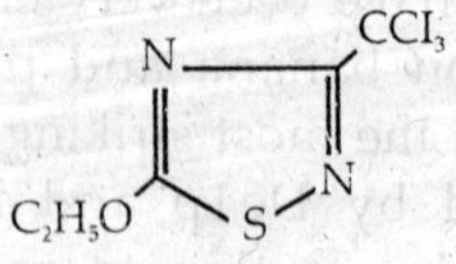

CCl$_3$
N
N
C$_2$H$_5$O S

Terrazole

(iii) *Thiophanares*

Thiophanates seem to occupy an exceptional position since it is not a heterocyclic compound, but instead is a disubstituted benzene derivative. In 1962, Aelbers related the development of these bisthioureido benzene derivatives to the discovery by Sijpestein and Pluijgers of the systemic protectant action of phenylthiourea. It was found that both methyl and

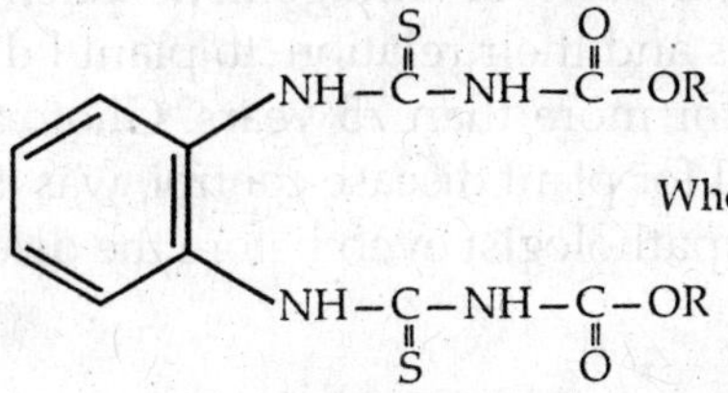

Where $R=CH_3$ OR C_2H_5

Thiophanate

ethyl thiophanate are broad spectra fungicides ,which have each the same antifungal spectrum as benomyl, though it is on a somewhat lower level of activity.

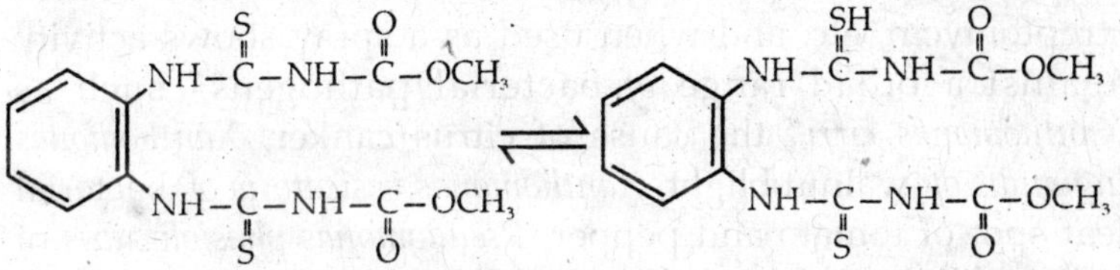

Methyl thiophanate

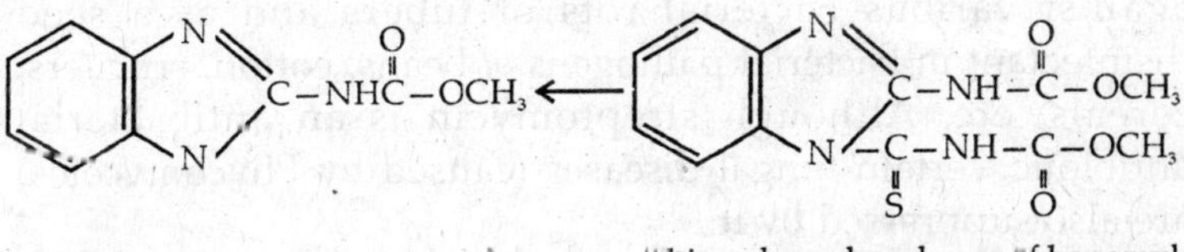

MBC

Thiocarbomyl analogue of benomyl

Antibiotics

Antibiotics are chemicals which are produced by micro-organisms and act again micro-organisms. Most antibiotics known until now are products of actinomycetes and some are from fungi and bacteria. The chemical nature of antibiotics is complex and is not, as a rule, related to each other. A large number, of antibiotics have been tried for plant disease control but only a few have been successfully utilized. The important antibiotics for t the control of plant diseases are streptomyčin, tetracyclines, griseofulvin, cycloheximide and aureofungin. The use of antibiotics is a comparatively new method of plant disease control, but antibiosis or antagonistic phenomena among micro-organisms and their relation 'to plant I diseases have been recognized for more than 75 years. Gliotoxin, the first antibiotic to be used for plant disease control, was isolated and purified by a plant pathologist even before the discovery, of penicillin.

(i) Streptomycin

It is produced by the actinomycete *Streptomyces griseus.* Streptomycin controls both gram positive and gram negative plant pathogenic bacteria. Streptomycin or streptomycin sulphate is sold as agrimycin, phytomycin, ortho-streptomycin, etc. and when used as a spray shows activity against a broad range *of* bacterial pathogens,' such as *Xanthomonas cirri"* the cause of citrus canker, *Xanthomonas juglandis of* walnut blight, *Xanthomonas vesicatoria of* bacterial leaf spot of tomato and pepper, *Pseudomonas phasioloculus* of hollow blight *of* French bean, and *Erwinia amylovora* of the fire blight of apples and pears.

Streptomycin is also used as a dip for potato seed pieces against various bacterial rots of tubers and as a seed disinfectant in bacterial pathogens *of* beans, cotton, crucifers, cereals, etc. Although streptomycin is an antibacterial antibiotic, certain fungal diseases (caused by Phycomycetes) are also suppressed by it.

(ii) *Cycloheximide*

This is commonly referred to as 'Actidione'-a trade name *of* Upjohn Co.—and is obtained as a by-product in streptomycin manufacture. Cycloheximide (Actidione). a glutarimide antibiotic, is produced by Different species *of Streptomyces.* including S. *griseus* and S. *nouresi.* A demonstration by Kerridge, showing that the cycloheximide is a potent inhibitor of protein synthesis in *Saccharomyces carlsbergensis,* was followed by similar observations involving other organisms by various workers. It is active against a wide range of fungi and yeasts but is inactive against bacteria. Cyclohoximide is the common name for β-[2-(3,5 dirnethyl-2-oxcylohexy1)-2-hydroxyethyl] glutarimide.

(iii) *Griseofulvin*

This is produced by *Penicillium griseofulvum*, P. *patulum*, P. *digricans* and P. *jancyewski* is toxic to several plant pathogenic fungi, such as members of powdery mildews, *Botrytis fabae, Alternaria solani* and some rusts. It is active against chitinous walled fungi. Griseo. fulvin has been successfully used *for* the control *of* tulip fire caused by *Botrytis tulipae.* Root application *of* griseofulvin was found to be quite effective against certain powdery I mildews, *Botrytis* diseases and bean rust *(Uromyces appendiculatus).* The systemic control *of* several powdery mildews which are rather resistant to most fungicides is interesting.

Brian concluded that systemic control is not very practical since it is usually found to be no better *or* worse than other fungicides already in general use. In 1960 Brian postulated a hypothesis that griseofulvin interfered with the biosynthesis *of* chitinous elements *of* fungal cell walls, which could also explain the relative non-toxicity of this chemical to animals and higher plants. Subsequent studies have shown that this antibiotic does not inhibit the synthesis of protein, chitin and lipids, nor does it alter the rates of respiration, glycolysis *or* cell well synthesis in fungal cells. Byrde reported that griseofulvin (0.1%) Is effective as a foliar spray for the control of apple rot caused by *Sclerotinia fructigena. Monilia* disease *of* apples caused by S. *mali* was successfully, controlled in Japan by griseofulvin. For further information on the use *of* griseofulvin in agriculture, see Ford *et al.* (1958).

(iv) *Blasticidins*

It is produced by *Streptomyces griseochromogenes.* In Japan it has been utilized for the control *of* blast disease of rice, *(Piricularia oryzae).* It is active' against both bacteria and fungi but is selective in its activity against the latter. The antibiotic inhibits the rice blast fungus at concentrations *of* 5 to 10 μ g/m!. This antibiotic interferes with protein synthesis.

(v) *Aureofungin*

Aureofungin is a new broad spectrum antifungal antibiotic produced in submerged culture by *Streptoverticillium cinnamomeum* val. *terricola* Thirum. Chemical characterization indicated that it belongs to a new aromatic group of heptanes and is, therefore, distinct. Its remarkable property, is that it is absorbed when applied as a spray *or* given to the roots as a solution. Thus, it is translocated and can be detected in other parts of the plant. This has been detected both by bioassay methods as well as by spectrophotometry.

The most useful property *of* the antibiotic is its high activity against a large number of phytopathogens and its absorption and translation in living plants. Some of the examples of effective control of plant diseases incited by pathogenic fungi are: *Citrus* gummosis incited by *Phytophthora* species, identical with *P. citrophthora,* which is a devastating disease in several parts of India and which has been controlled by aureofungin and powdery mildew of apples incited by *Podosphaera leucotricha* which is completely controlled by spraying 100 ppm antibiotic spray solution.

Downy mildew, powdery mildew, and anthracnose or-grapes are controlled by spraying aureofungin . Seed treatment of rice effectively controls *Helminthosporium oryzae.* Similarly, *Pyricularia oryzae* of rice and ragi are effectively controlled by aureofungin. *Diplodia* rot of mango and *Alternaria* rot of tomato, *Sclero tinia or* rot of peach .

(vi) *Tetracyclines*

Tetracyclines are antibiotics produced by a number of species of *Streptomyces.* Terramycin, aureomycin and actinomycin have been used for plant disease control. Oxytetracyline has been used as a soil-drench or as root dip for controlling crown gall and has been utilized for the control of fire blight of apples and pears in conjunction with streptomycin.

Tetracycline has recently been found to be effective against plant diseases caused by, mycoplasmas. An injection of a concentrated solution of tetracycline or oxytetracycline into tree trunks offers possibilities, of curing peach-x, pear decline coconut palm yellows and citrus greening .

(vii) *Agrimycin-100*

This is 15% streptomycin sulphate and 1-5% terramycin. Agrimycin-500 is 1.755% streptomycin sulphate, 0.176% terramycin and 42.4 % metallic copper. It has been utilized for the control of bacterial diseases, such as *Erwinia amylovora*, 'halo-blight caused by *Pseudomonas phaseolicola*, citrus canker caused by *Xanthomonas citri*, seedling blight, leaf spot and black arm disease of cotton caused by *Xanthomonas malvacearum* and soft rot and black leg of potato caused by *Erwinia carotovora.*

(viii) *Antimycin*

Antimycin, a macrolactone type of antibiotic, is produced by several species of *Streptomyces*, including S. *kitasawensis* and S. *griseus*. This antifungal antibiotic has been reported to be effective against early blight of tomato, seedling blight of oats caused by *Helminthosporium* and rice blast caused by *Pyricularia oryzae*. However, because of its phytotoxicity, its use is limited.

(ix) *Thiolutin*

This is a water soluble pyrothin-type antibiotic produced by *Streptomyces albus* It has been used on a limited scale in the control of late blight of potato and the downy mildew of broccoli caused by *Peronospora parasitica.*

(x) *Nystatin*

Nystatin is also known as mycostatin or fungicidin. It is a polyene (tetraene) antifungal antibiotic produced by *Streptomyces nourseL.It* is manufactured for clinical use but has also been used to a limited extent for plant disease control. This has been found efficacious in controlling anthracnose of beans and downy mildew of cucumber and as seed treatment against stripe disease of barley. Ananthanaryana and Seshadri reported that it can be utilized as post-harvest dip against peach brown rot and anthracnose of banana.

(xi) *Bulbiformin*

Bulbiformin is an antifungal polypeptide antibiotic produced by *Bacillus subtilis* and was reported by Vasuaeva. in the year 1958. The production of this antibiotic has been demonstrated in sterilized soil amended with molasses, groundnut cake, sweet clover roots, etc. Inoculation of amended soil with *B. subtillis* controlled the wilt disease caused by *F. oxysporum* f. sp. *udum* on pigeon pea. Singh demonstrated the production of bulbiformin on seed coats, spermatosphere and rhiiosphere by bacterized pigeon pea seed sown in sterilized andunsterilized soil. The antibiotic was absorbed, and because it is systemic, considerably reduced the incidence of wilt of pigeon pea.

(xii) *Pentaene G-8*

Batra and Bajaj isolated an anti fungal antibiotic from the mycelium of *Streptomyw anandii.* It belongs to the Eurocidin group of pentaene antibiotics and they designated it as pentaene G-8. The important fungal plant pathogens which are inhibited by this antibiotic are *Colletotrichum* spp., *Helminthosporium* spp. and *Puccinia* spp. Complete inhibition of chilli seed borne fungi has been obtained by the seed" treatment of S. *anandii* culture filtrates.

(xiii) *Endomycin*

This polyene antibiotic produced by *Streptomyces endus,* though non-toxic to higher plants, has been found to be effective in reducing the intensity of only a few diseases, such as leaf rust of wheat and fruit rot of strawberry caused by *Botrytis cinerea.*

Soil Fumigants

The most promising method of controlling nematodes in the field has been through the use of chemicals called nematocides. Some of these, including chloropicrin, methyl bromide, mylonet vapam and vorlex; give off gases after being applied to the soil. They are general purpose preplant fumigants and are effective against a wide range of soil micro-organisms, including, in addition to nematodes, many fungi, insects and weeds. These are also highly volatile fumigants, such as formalin and carbon-disulphide.

(a) *Formalin*

Various chemical treatments of the soil have been developed, one of the oldest being drenching with formaldehyde. Formaldehyde corrodes the hands and is somewhat erratic in effectiveness. For the control of damping off and seedling blights 37-40% solution in water is used. Excellent control of onion smut has been obtained in the-USA by a solution of formaldehyde (1-128 parts) ,applied at, seeding time at a rate of 200 gallons per acre with the help of machines especially devised for the purpose.

(b) *Carbon Disulphide* (CS_2)

Carbon disulphide is perhaps the earliest chemical used as a soil fumigant for checking nematodes. Carbon disulphide is used in citrus orchards against *Armillaria mellea.* It is used like formalin. This substance was originally employed in 1872 by Thenard against phylloxera of the vine *(Phylloxera vitifoliae* Fitch). Carbon disulphide is cheap, volatile and a insecticide.

(c) *Chloropicrin*

This is useful both as a fungicide and a larvicide. It is injected into the soil at a depth of 3 to 6" in holes 9-12" apart. The soil is covered with impervious cloth or plastic sheeting for 48 hours.

(d) *Vapam*

Other fumigants produced in recent years include methamsodium (Vapam: sodium Nmethyldithiocarbamate), a colourless liquid which decomposes rapidly in moist soil to release a fumigating gas.

$$H_3C-\underset{\mid}{\overset{}{N}}-\underset{\|}{C}-Na$$
$$\quad H \quad S$$

Vapam (SDMC)

Vapam is not volatile but decomposes in soil to yield methyl isothiocyanate (CH_3-N=C =9). In addition to methyl isothiocyanate. Vapam also yields carbon disulphide, hydrogen sulphide, methylamine and other products when it is mixed with the soil. Vapamhas been used as a nematicide and selective weedicide. But it has been found to, be quite effective in the control of wilt of cotton, damping off of papaya, root rot of beet.

(e) ***Methyl Bromide***

Methyl bromide has also been use a as a soil nematicide, but it is too phytotoxic to be used in fields with a standing crop. It should be applied in the soil as a fumigant a few weeks before the crop is sown or planted. The use of this sterilizing gas has been restricted to the treatment of nursery beds, vegetables and fruit production and ornamentals. Methyl bromide has low boiling point *(4.5°C)* and requires a plastic cover for effective confinement of the gas, or for application with deep injectors.

(f) ***DD Mixture***

Dichloropropene and dichloropropane (DD mixture) and ethylene dibromide (EDB) are two effective nematicides which came into the market soon after World War II 'and are used extensively for nematode control. Dichloropropene was first introduced by Carter in Hawaii for the control of soil-born diseases in pineapples. Carter's material, consisted of a mixture of 1,2 dichloropropene and 1,2- dichloropropane.

(g) ***Dazomet***

Dazomet (tetrahydro 3,5-dimethyl-2H-1,3,5-thiadiazene-2-thione) is a wettable powder used at about 440 kg/ha and is, applied with a fertilizer spreader or by spraying suspension in water, followed by thorough mixing and 'sealing in' with water. In the soil, dazomet' releases methyl isothiocyanate and is effective for the control Of nematodes, wireworms and some soil-borne fungi, such as Pythium, *Rhizoctonia, Fusarium, Verticillium* and *Colletotr; chum* (Martin; 1968).

Oils

Various types of oils are 'capable of controlling several plant diseases. The effectiveness of oils in the control of plant diseases has been reviewed by Calpouzos. Shigatoka disease of bananas caused by *Mycosphaerella musicola* has been controlled by oils, which, because of their advantages have almost completely replaced conventional fungicides. Oil sprays are now widely used *for* the control of *M. musicola.* Nearly 50 million litres are estimated to have been used in Latin America and the Caribbean in 1964 and perhaps another 7.6 million litres in other banana producing areas.

The advantages of oils are

(i) superior disease control,

(ii) effectiveness at very low dosages of even less than one gallon per acre,

(iii) excellent spreading and sticking properties on leaf surface

(iv) low cost, and

(v) little or no toxicity to man and animals.

There are several disadvantages also. These are

(i) effective against a limited number of plant diseases, and

(ii) phytotoxicity may occur on sensitive plant species.

The oils involved in disease controlled mainly mineral oils from petroleum, and to a much lesser extent, glyceride oils from plants, 'and synthetic oils.

Oils have been used successfully in the control of greasy spot of citrus caused by *Cercospora citri-grisae.* Oil alone controls angular leaf spot of tung tree caused by *Mycosphaerella aleuritidis.* Oil sprays have also been successfully utilized for the control of *Septoria apii* on celery.

Safeners, Spreaders and Stickers

Safeners

A safener is defined as a chemical that reduces the phytotoxicity of another chemical.

Copper sulphate is phytotoxic, but with the addition of lime its toxicity is reduced. Lime is, therefore, a safener. Lime is used universally with chemicals to prevent the formation of, or to neutralize arsenic acid, which is phytotoxic to sensitive foliage. glyceride oils act as good safeners for copper sprays by providing a film around each particle, but they reduce the fungicidal value of the product.

Spreaders

Materials which are added to the spray mixture to establish improved contact between the fungicides and the sprayed surface are known as spreaders.

These can be classified as water-soluble and water-insoluble. Mineral oils, glyceride oils and terpene oils are water-insoluble spreaders. Soap has long been known as spreader. Sulphated alcohols,sopamines, resinates and petroleum sulphonic acids are some of the common spreaders available in the market. Spreaders reduce the surface tension and improve contact between the fluid and the sprayed surface so that the fluid does not collect large drops and run off easily.

Stickers

Materials added to spray or (lust which improve its adherence to plant surfaces are known as stickers.

Flour, starch, gum arabic, and dextrins are common used as stickers. Bentonite clays when add to dusts improve the latter sticking quality. Oils, oil emulsions and hydrocarbon oils have also been employed as stickers.

Carriers

When the fungicide is used as a dust, the active ingredient generally diluted with a finaly divided powder, called a carrier.

Origin of Fungicide Resistant Strains

The disappearance of resistance in the absence of the fungicide could be due to sublethal concentrations of the latter. This results in the formation of a substance or substances which confer resistance which would otherwise not be formed. Perhaps, adaptive enzymes, the formation of which is induced by the fungicide, are involved.

Stable resistant strains perhaps arise by mutation which occurs either by chance, or in the case of certain fungicides, is possibly brought about by the fungicide itself. It is significant that the exposure of *Fusarium caeruleum* (potato dry rot) to tetrachloronitrobenzene (TCNB) resulted in the appearance of TCNB resistant mutants and that resistant variants of *Botrytis allii* (onion neck rot) were produced after the fungus was treated with the vapour of pentachloronitrobenzene. Some aromatic hydrocarbons are said to produce gene mutation in fungi. The fungicide resistance of some fungi, including a few which attack plants, has been shown by genetical analysis to be of chromosmol origin, as in *Hypomyces solani* f. sp. *cucurbitae* (seedling blight and fruit rot of cucurbits), where resistance to TCNB is determined by three independently inherited genes

PREPARATION AND METHODS FOR USE OF FUNGICIDES

Somers has discussed the various aspects of fungicide formulations. Formulation of fungicide is mainly concerned with methods of presenting the active ingredient in the most effective physical form-effective with regard to storage, application and ultimately the biological activity. A good fungicide is likely to have the characteristics mentioned below:

a. Fool-proof	e. Easy spreading nature
b. Non-toxic to host	f. Easy to apply
c. Easily available	g. Economical
d. long residual effect	h. Simple to prepare.

The selection of a fungicide, manner and mode of application are important considerations that determine the success of operation. The efficient use of any fungicide success of operation. The efficient use of any fungicide depends largely on dispensing the active substance to the right place at the right time. Biochemically the fungicides are very active and only small amounts are needed at the relevant surface. This property affects the application technology causing considerable problems in obtaining a uniform distribution of fungicides on non-uniform targets. Fungicides may be diluted either with a solid and applied as dust or with a liquid and sprayed on the plant target. Water is a cheap and effective dilution medium, easily available and with a few exceptions, used as a carrier for agricultural sprays. Fungicides have a low water solubility and hence are treated to make them compatible with water. Surface-active ingredients are needed to prepare water dispersible powders WD, stock emulsions, or emulsifiable concentrates EC. Moreover, surface-active ingredients are required for improving the spreading, suspending and wetting properties of the spray. To improve the weather resistance of deposits, stickers may also be incorporated with the spray. Stickers when, incorporated also improve spray deposition and penetration. Dusts are often the simplest and cheapest way of applying a pesticide, but climatic conditions mainly limit their use as compared to sprays.

Types of Formulations

Dusts

As seen by the the name these formulations are applied in dry form as dusts. The formulations generally contain less precentage (<10%) of the active ingredient. The other component known as diluents or carriers function to improve the application properties of the formulation and to dilute economically a relatively expensive chemical. Dusts are prepared in two stages. Firstly, a dust concentrate containing more than 25% active ingredient is formed by mixing the fungicide with a highly sorptive diluent by impregnation or grinding techniques. This dust concentrate is convenient in transportation and storage, before the final blending and dilution to a field strength dust is done. The fungicide is first mixed with a carrier and later diluted with a diluent.

An important determinant of dustability is the particle size. Well-formulated dusts are so manufactured that they are light enough to be carried by a slight breeze for a considerable but controlled distance, possess good crop covering properties and form an adherent deposit on the plant surface. Formulations should not become sticky under humid conditions. The most widely used diluents are the silicate minerals attapulgite, kaolin, talc; bentonite calcium silicate, hydrated silica, diatomaceous earths, calcium and magnesium carbonate, gypsum, lime, etc. The use of granular formulations of pesticides has developed rapidly in recent year. Due to their size, the granules do not drift when applied, as do dusts with fungicides. Granules have limited application, being confined to soil and seed treatments.

Wettable and Water-dispersible Powders

This is the most commonly used method. Many fungicides, especially the agricultural fungicide spray are formulated in this form. Modern wettable powders have the quality to wet easily and disperse well in water. Spread stickers are desirable for plant surfaces with glossy or waxy surfaces in addition to a wetting agent. Wettable powders usually contain 25 to 50% active ingredients. The suspending agent with formulation has two functions: to deflocculate , the individual particles so that they form a smooth dispersion, and acts as a protective colloid which stabilizes the dispersing system and reduces sedimentation in the spray tank. Materials such as sodium lignin sulphonate (sulphite lye), methyl celluloses, poly-vinyl acetate, aluminium silicate are used as suspending agents. A highly developed type of water dispersible powder is called a colloidal powder in which the fungicide is so finely divided that individual particles never sediment out.

Emulsions

In such formulations the fungicide is mixed in a water-immiscible solvent such as a petroleum fraction. With the aid of a surface active agent this can be dissolved in water to form a stable emulsion. The solvent should be non-phytotoxic.

Formerly, stock emulsions stabilized by such hydrophilic colloids as CMC or gelatin were prepared but now these have been largely replaced by emulsifiable concentrates EC or water dispersible oil. EC need a high concentration of oil soluble emulsifying agent, often more than 5% of their total weight, to give instantance formation of a homogeneous emulsion on dilution in water. At these concentrations the highly surface active emulsifiers confer good wetting properties to the spray.

The efficiency of an emulsifier is dependent on the balance between the hydrophobic group of the molecule in the oil phase and the hydrophilic portion in the aqueous phase. This has led to the concept of the hydrophile lipophile balance which can be used as a guide in choosing an emulsifier for a specific formulation.

Developments in spray machinery have made it possible to apply invert emulsions, that is, water in oil, which are less susceptible to drift than normal emulsions.

From the control point of view plant diseases can be grouped under four heads:

(a) soil borne,

(b) seed-borne,

(c) air-borne and

(d) insect borne.

Thus, the time and method of application of fungicide for plant disease- control depends on the nature of the disease.

Soil-Treatment

This is avoid method of soil disinfection and is primarily aimed at killing the soil pathogens and making the soil safe for plant growth. This treatment can be given in the various forms which are described below

Fumigation

This by definition means treating the soil with chemical fumes for killing pests or sterilizing soil. True soil fumigants are highly volatile chemicals such as carbon disulphide, chloropicrin, and methyl bromide. The method of application varies according to the chemical' used and area of soil to be treated.

Trench Method It is a simple method consisting of a glass jar fitted with a lid in which holes on opposite edges have been punched, The fumigant is placed inside the jar and poured into the bottom of trenches at a depth of 6 to 8 inches in the area of the soil to be treated. Trenches spaced 9 inches apart are closed soon after the application of the chemical. The chemicals are also applied at the point of line intersection in a grid design established on the soil surface. The chemical in this modified trench method is poured into the holes in that soil through a funnel to the desired depth.

Multiple Point Application A hand hold applicator is designed for single point application and such a device has a reservoir tank, depth gauge, injection pump, delivery tube and metering device to vary the amount of material applied. The applicator is held by a T-shaped handle and pushed into the soil to the collar at the base-the depth gauge. The material' is then pumped out through the delivery tube into the soil.

Application pattern is established on the surface by a grid system with intersecting lines on 6 x 6- or 9 X 9-inch centres. The amount of material that is applied in mm per injection determines gallonage to be applied per acre. The maximum space allowed for effective control of the soil-borne plant pathogen largely depends on the material used, dosage applied, depth of application and edaphic factors. Single point application is generally adapted in greenhouse operation where individual soil containers for individual plants are treated, and in field testings, this permits the establishment of maximum spacing for multiple point application. This will more or less permit the continuous treatment of large areas of soil.

Once a fumigant is applied to the soil, some form of surface seal is usually attempted, and in such applications, seatings can be done all the point of application.

Gaseous Fumigants Highly volatile gases, for example, methyl bromide, are packed in sized containers appropriate for various fumigation practices. The plastic covers of the containers must be kept above the soil surface which has been treated in the fumigant. Tractor-mounted equipment used for fumigation can include attachment to lay a plastic ground cover, and to move soil over the edge of this cover to make an almost gas tight seal.

Surface-Treatment

The appropriate method of applying fungicide to the soil depends on various factors:

a. Chemical and physical properties of a compound.

b. Relative activity of the compound against pathogen.

c. Nature and ecology of the target organism.

d. Tolerance of the organism to the fungicide.

e. Cost of the compound and nature of formation.

Surface Application

Fungicides are applied to the soil surface as dusts ; as wettable powders or emulsions (using standard spraying equipment); as granules (with granular applicators); or with the me of fertilizer spreaders. Fungicides may be applied by aerial applications or after mixing with the fertilizer. A good example of surface application is the use of Brassicol (PCNB) for the control of southern blight on peanuts. There has been a considerable difference in the performance of various PCNB formulations. At equivalent rates the formulations which result in better distribution of the, fungicide on the soil surface (e.g. sprays) generally give better control than granular formulations which leave portions of the soil surface untreated. Such an application is more effective where the pathogen is located in the upper layers of the soil and not throughout the soil.

Surface Application Followed by Mixing

Fungicides must be mixed in the soil where the pathogen is found throughout the-soil profile, when the nature *of* the fungicide does not permit its presence on the Boil surface, *or* when a systemic fungicide is used.

In-Furrow Applications

These types *of* applications are specifically designed to control seedling diseases., Four techniques can be considered under this head:

A) *Dust Applications.* Machinery utilized for such application consists *of* a tractor mounted hopper with a powered slow moving agitator (*for* granules) *or* rotary drum with air hoses. The equipment is calibrated for the accurate delivery *of* desired amounts *of* fungicides to the seed furrow. Dusts *or* small granules can be blown away, but in modern dusters, such losses have been reduced by: (a) incorporating multiple outlets *for* dust dispersion in the round cover at the delivery end *of* the discharge tubes, (b) high air velocity in the tubes to maintain the dust in suspension and (c) low air velocity at the discharge point *to* eliminate *losses* due *to* drift.

B) *Spray Application.* This is a relatively old method *of* fungicide application, used mostly for the control *of* seedling diseases. The equipment consists *of* a mixing tank where the fungicide emulsion *or* disposition is placed and agitated, a pump which gets power from the tractor power take off or independently' from a gasoline engine; a manometer, a control valve, a by-pass line to return *to* the tank and a manifold with a number *of* outlets with an outlet for the treatment *of* each row. The number *of* ,nozzles used *to* spray a row as well as their placement 'depends on the crop. The nozzle with a conetype spray pattern is placed between the seed press wheel and the closing disc *of* the planter.

C) *Planter Box Dust Applicators.* This is a simple method in which seed and fungicide can be delivered simultaneously without additional equipment. Dust or wettable powder formulations are applied simultaneously with the seed and mixed in the planter box *or* hopper.

The seed which is partly coated falls, together with some fungicide,' into the seed furrow in the planting operation, and it is thus easily protected. This method does not' uniformly apply fungicide to the soil and to-an types *of* seeds, and hence, its simplicity is the only advantage.

D) *Dips and Drenches.* In this method, seedlings are dipped in transplant water containing fungicide *or* applied as drench the fungicide suspension in water *to* the soil in seedling beds. The limiting factor is the large volume *of* water required and the cost *of* transportation. But with systemic development interest has again been diverted to such applications.

The use *of* seed-treatment fungicides and their application has been discussed by Purdy (1967) and Kabana *et al.* (1977).

Seed-treatment

A number of seed-borne diseases are amenable to control when the proper chemical is applied to the seed.

Systemic fungicides have been found to be successful in removing systemic and deep seated infections. Contact fungicides (thiram and captan) usually effective against a broad spectrum fungi, are still widely accepted. Certain seed-treating devices are discussed below.

a) Batch-type Treaters

A shovel is the simplest form of such a treater. The method involves placing the seeds on a floor, scattering the recommended amount of dust over it, putting another layer of seeds on it, spreading the recommended dose of fungicide on it, and then mixing these till no streaks of the dusting is visible. Only volatile fungicides are added in this way.

b) Gravity Treaters

These treaters are used for applying chemicals to small grain seeds. One type of this treater consists of two sets of cones over which the seed and dust fungicides are passed and mixed. The upper cone of each set is perforated to allow dust to pass through. The point of the cone faces upwards to let the cone function as a seed divider. As the seed is passed over the perforated cone, it is collected in the lower cone and mixed with the fungicide. The seeds after passing through all the sets of cones are discharged at the base of the treater and bagged or placed in bulk storage.

c) Continuous Flow Dust Treaters

The main feature of this machine is that it can meter the seed into the mixing chamber and also meter the copper carbonate formulation and bring it in contact with the seed. The seeds and the dust are mixed in various types of chambers that tumbles the seed in a rotating drum.

d) Slurry Treaters

Fungicide formulations for slurry applications containing wetting agents are referred to as wettable powders. In this Method, wettable powder is placed in a water carrier to form a thick suspension. Generally, all slurry machines have an adjustable hopper to control the flow of seed into the machine, a slurry tank with an agitator to keep the wettable powder in suspension, positive seed and slurry metering devices, and an auger to mix the seed and fungicide and move it to the discharge spout of the treater.

e) Drum Treaters

Drum type seed treaters came into use with the introduction of liquid formulations of organic mercurials. A drum treater is like a slurry machine with an adjustable hopper to control seed flow, a positive seed and fungicide metering device, and an inclined drum with baffles at the discharge end to control the time of seed retainment inside the rotating drum mixing chamber. The fungicide flows .down the outside of the fingers as the seed enters the mixing chamber through tubing attached to metal fingers at the inlet end of the drum. The fungicide is then distributed over the remaining seed by the tumbling action of the latter in the mixing chambers.

f) Mist Treaters

Even with the development of a drum type treater, coverage of the seed was not complete and a more refined machinery was required. The mist treater is the only machine so far developed that has apparently solved the problem of complete coverage. The machine consists of an adjustable hopper to regulate the seed flow, a positive seed and fungicide metering device, and a rapidly spinning disc that breaks the fungicide solution into small droplets. The seed is allowed to pass through the zone of five droplets of fungicides (mist) under the action of gravity and is then discharged from the machine.

g) Spray Treaters

This type of treater makes use of spray nozzles to mix the fungicide suspension with the seed in the mixing chamber. A continuous type of spray treater has an adjustable hopper to control the flow of seed to the chamber, a tank containing the treating solution or suspension that is equipped with a signal device to show when the mechanical agitator is operating, pressure gauges showing the pressure in kg/sq. cm at the tank and nozzle ends *of* the pressure system, one fiat fan nozzle, a monitoring system *by* which the flow of the treating solution or suspension is controlled, a rotating cylinder about 2 metre long and big enough to permit good distribution of the fungicides, and a corrugated liner for the mixing drum. The treating suspension is placed under pressure in the tank, passes through lines of tubing to the nozzle and is sprayed on to the seed. The fungicide is redistributed *by* the tumbling action of the seed in the mixing drum.

Compressed air provided *by* an air compressor aids the distribution of powdered chemicals when introduced into the mixing drum.

Hot and Cold Water-treatment

This treatment is very important for the control of certain seed-borne diseases, and the apparatus for cold soak treatment can be a drum, a barrel, or a similar container *of* sufficient size to be filled to three quarters capacity with the seed to be treated. A device a plastic sheet-is needed to seal the container after the soaking solution has been drained from the container. After the treatment is over, facilities for drying the seed are needed.

Selection of seed lots by a "vigor rating" system is being accepted in the seed industry. This system relies on a quantitative measure of growth al well as an evaluation of germination. The vigor index developed allows the selection of seed lots and fungicide treatments that give best performance of both the variables.

Seed-treatment fungicides when applied to seeds before planting are in the form of liquid or liquid slurries.

Amongst the most destructive enemies of mankind are rusts of wheat.

The most common examples of foliar diseases are: (a) disastrous famine in Ireland in the middle of the 19th century due to late blight infections of potatoes; (b) epidemics of powdery mildew of vines, (c) apple scab, (d) rusts and (e) smuts.

Aerial diseases cause losses in the quantity and quality of field. According to their function, fungicides can by classified as:

a. *Protectollts-applied* to protect the plant from expected infection.

b. *Therapeutants-applied* soon after, the onset of infection, they suppress the development of the pathogen and thus cure the plant.

c. *Eradicants-applied* to diseased organs to destroy the inoculum or prevent its formation.

Methodology of Fungicide Application

Millardet applied Bordeaux mixture to the levels of vines by means of a brush made of heath straw. At that time application of fungicide to plant foliage was a breakthrough in disease control. For the control of foliar disease either the fungicides are applied topically direct to the foliage, indirectly to the soil or into the trunks of the trees (in the case of systemic-toxicants). Each method requires a specific suitable formulation. Many of the spray applications utilize water as a carrier, with formulation aids, such as solvents, emulsifiers, dispersants, wetting agents, stickers, stabilizers and safeners, added to the active ingredient by the manufacturer.

Spraying of Foliage and Fruit

Spraying techniques classified according to the volume of spray applied per unit area, are discussed here. The wetting of the foliage, and subsequently, the density of coverage affected by the volume sprayed, are important.

a. *High volume-more* than 400 litres/ha.

b. *Low volume* - 5-400 litres/ha.

c. *Ultralow volume-less* than 5 litres/ha.

High Volume Spraying

An the foliage and fruits are thoroughly sprayed to cause run off. The thickness of the film deposited is related to the interaction of the characteristics of both the formulated toxicants and the nature of the plants surface, and depends largely on the angle of the tilt of the surface. Usually the fungicide drained in the run off process is wasted and this loss may be 50% or even 90%. A general characteristic of run off spraying is the accumulation of deposit at the tips and edges of the leaves and fruits.

Low Volume Spraying

The spray in this method is distributed on the surface as discrete droplets and no run off occurs. Labour required to spray a crop unit is saved and the dose of fungicide is often reduced by 20-25%. In practice, the spraying equipment' emits droplets of different sizes. Low volume spraying ,by an aircraft has several specific advantages including that of reduced cost.

Ultralow Volume Spraying

This method, although similar to low volume spraying, has several advantages, and also lowers the cost of labour per unit of the treated area. The main disadvantage is that when using very fine spray, a wind velocity of more than 5m/sec greatly increases the drift hazard.

Application by "Sprinklers"

In such an application the chemical is introduced into the irrigation system and sprinkled on the foliage. Although thorough coverage is not "expected, redistribution may correct this handicap. The wastage of fungicides falling on the ground cannot be avoided. This method can be applied at the end of an irrigation cycle, when ground sprayers cannot be operated.

Application by "Netting"

The netting method was experimentally (devised in Japan for treating fruit trees. Fungicide is placed at the top of a tree canopy and rain helps to redistribute it over the whole foliage. In this method, fungicide is impregnated in the fibres of a net and the net is placed on top of the tree.

Foliar "Dusting"

Fungicidal dusting covers the entire foliage. However, its use is limited due to its weak tenacity. Dusting is carried out by emitting a blast of air in which the dust particles are borne. In the case of hand dusters the air is pumped by a plunger and the compression stroke of the piston expels a blast of dust laden air. Knapsack dusters are operated by a bellows, a hand lever operates the bellows and agitates the screens to allow the dust to flow into a hose and a pipe tipped with a sieve type nozzle. Power dusters driven by motors may have along boom with many delivery pipes. Sometimes a hood is attached to the boom to reduce, the amount lost by drift. Dustings can be accomplished by aircrafts also.

Application by "Vaporization"

This method is applicable in enclosed chambers. The

fungicide is thermally vaporized, usually by an electric heating element, and hot vapours condense due to the temperature difference to form minute particles which are dispersed in the air and which finally settle on plant surfaces. Usually the treatment is made at the end of the day in the greenhouse when all the ventilation systems are closed. This method is suitable for chemicals which are thermostable, for example, *sulphur* which vaporizes at 112°C and is generally used to control powdery mildew. *Ch/orothalonil* is available as tablets for thermal vaporization.

BREEDING FOR DISEASE RESISTANCE

The idea of resistance to pests and diseases has been a main centre of attraction for scientists, plant breeders ever since the demonstration near the turn of the century that resistance could be selected and that it is often simply inherited. The built-in protective resistance effective throughout a crop' plant life offers a compelling alternative to fungicides and insecticides. In the early years, minimizing the use of protective chemicals was desirable because it saved money but today there is an added advantage of reducing their hazard against those pests and diseases such as cereal rusts, soil-borne smuts and certain nematode diseases.

The use of disease resistant varieties for controlling plant diseases has been coined as the "painless method" because it does not cost the farmers anything. In an underdeveloped country like India, it is all the more because of the heavy costs of spraying and dusting crops on a large scale. The use of resistant cultivars and hybrids has several other advantages. There is no hazard to human health and wild life which is caused by large scale use of dangerous fungicides and pesticides. It also decrease pollution which results from the use of poisonous chemicals and their residues. Resistant crop varieties check epidemics of pathogens and pests and thus help to maintain the biological balance in the environment.

Resistance can be varied in different hosts and parasites. If can be mechanical, such as waxiness, hairiness and toughness of the epidermis or it may be biochemical in which case it is the result of the interaction between the cytoplasm of the host and the parasite. The biochemical defence mechanism is due to the synthesis of poly, phenols in the host that may be toxic to any organism present and to the plant itself. Susceptible and resistant reactions differ principally in the speed and scale of host responses which are more rapid in the latter. As a result, the development of the invading parasite is restricted.

Susceptibility and resistance to a parasite are in a large measure inherited characteristics. The laws governing their inheritance are the same as those governing other characteristics of the host and parasite. It has been proved in several cases that the resistance or susceptibility to disease is located in the genes. The breeding of resistant varieties of crops is, there. fore, based upon the laws of inheritance. The Darwinian principles of variations, struggle for existence and survival of the fittest apply to all the wild varieties and their pathogens.

Wild plants have been growing for centuries in the presence of the pathogens and natural selection must have eliminated the most susceptible biotypes and left the most resistant ones. In this way regional biological balances exist between the native plants and their pathogens and pests. The balance will tend to persist unless disturbed by new factors such as the introduction of new biotypes of pathogens or hosts. This is less likely to occur if the region is protected by barriers such as oceans or high mountains. Often it is man who disturbs the natural balance by transporting plants and pathogens to new regions, or who, through crop improvement, produces new varieties of crop plants. Many of the serious outbreaks of plant diseases are traceable to the introduction of parasites to which non-resistant crop plants immediately succumb.

Failures accompany successes while breeding resistant varieties. Plant breeders even now introduce genes for resistance. Borlaug has remarked that the average useful life of a rust resistant variety is as short as five years in some parts of the world after which it succumbs to a new race of rust. Walker wrote, "By and large the development of resistant varieties must be looked upon as a continuing programme. The potential variability of most pathogens (including viruses) will not permit any currently successful variety to remain so for an indefinite period". Several other examples can be cited where resistant varieties did not stay resistant for long, such as Hessian fly on wheat, powdery mildew on barley, *Cladosporium* leaf mold of tobacco, mosaic virus of tomato, golden nematode and leaf blight of porato. The question arises as to why these varieties were so short-lived and why the long and continued series of efforts failed to give the desired results. The reason is that while the breeder tries to make plants more resistant, nature tries to make pathogens more virulent, and ultimately nature gets the upper hand.

It was earlier stated that the characteristics of an organism are governed by the laws of genetics which are common to both the host and parasite. Resistance or susceptibility (both of which are relative expressions) is the result of interaction between these characteristics. It has now been shown that the genotypes of the host as well as the parasite can be varied by segregation-recombination of genes, mutation, heterokaryosis, somatic hybridization, etc. Genetical studies have now been extended to bacteria and viruses where new strains arise due to hybridization, transformation and transduction. The previous conception that the pathogen is a fixed entity which damages the crops year after year without change is no longer acceptable. There are a great array of genotypes of the parasite as there are of the host.

Type of Resistance.

Plant pathologists commonly classify resistance into monogenic, polygenic and oligogenic, according to whether resistance is governed by one gene, many genes or a few genes. Resistance can be described as genetical, mechanical or epidemiological.

The general behaviour or the genetic behaviour of disease resistance is governed in many simple to complex ways.

Monogenic resistance (governed by a single gene) is governed by one gene with 3:1 Mendelian ratio. It can be monohybrid dominant, such as the resistance of oats to *Helminthosporium victoriae,* of sorghums to *Periconia circinata,* of cabbage to the pathogen *Fusarium oxysporum* f. sp. *lycopersici,* of onion to the pathogen, *Pyrenochaeta terrestris* and to *Cercospora* leaf spot of cucumber. It can be monohybrid recessive, with resistance as a recessive character, such as powdery mildew of barley, sorghum smut and yellow rust of wheat.

Oligogenic resistance (governed by several genes) is controlled by two to several genes. Thus, it can be dihybrid, trihybrid, etc. Anthracnose of beans is governed by three genes.

Polygenic resistance (governed by many genes) involves many genes which are more difficult to analyze and -which are known to be involved in a large number of diseases such as cotton wilt.

Extrachromosomal Inheritance

Extrachromosomal inheritance of disease resistance is rare but is of great importance. The widespread epiphytotic of southern leaf blight of corn in the USA was due to the sudden appearance and distribution of a new race identified as race- T (Smith and Hooker, 1970) of *Helminthosporium maydis.* Resistance to the new race is mostly extra chromosomal. Corn with *cms-T* cytoplasm is more susceptible to yellow leaf blight caused by *Phyllosticta* sp. Symptom production in potato virus X in peppers is apparently determined by cytoplasm.

Gene Interaction

Disease resistance to pathogens is not always inherited. In many instances the Pattern of inheritance is quite complicated. The resistant phenotype is the ultimate expression of a chain of events in which genes for resistance perhaps control only a few links. Although genes may appear to act as autonomous units, in effect gene products interrupt and this becomes dear in genetic analysis by the appearance of complementation, modification and epistasis.

Modifier Genes

A host may carry a gene for strong resistance but this is not expressed because of the presence of an inhibitor gene. In onions, resistance to smudge' disease is controlled by a single dominant gene R. In the presence of a partially dominant inhibitor gene I the resistance is reduced or completely suppressed, depending on whether the gene I is present in the heterozygous or homozygous state. In wheat rust, resistance is completely suppressed by a dominant inhibitor gene.

Reversal of Dominance

A gene may be dominant for resistance to one race of the pathogen, but recessive Cor resistance to another race of the same pathogen. This reversal of dominance may be due to dosage effect, that is, one dose of the gene in a heterozygote being sufficient to provide resistance against one race of the pathogen but not against the other, or due to closely linked genes for resistance, one dominant and the other recessive.

Vertical and Horizontal Resistance

When a variety is more resistant to some races than to others, resistance is called "vertical", "perpendicular" or "specific". When resistance is such that it is uniformly spread against all races of the pathogen it is called "horizontal", "lateral", "generalized", "non-specific" or "field resistance". The derivation of this term is mathematical and refers to the horizontal axis of the graph. Horizontal resistance can be demonstrated when a series of pathotypes is tested against a series of pathodemes. Horizontal resistance has been called field resistance against *Phytophthora infestans* and generalized resistance against *Puccinia graminis.* Vertical resistance (VR) confers complete but impermanent protection, whereas horizontal resistance (HR) confers incomplete but permanent protection.

The concept of vertical and horizontal resistance (VR and HR) was put forward by Van der Plank. Vertical resistance is conditioned by oligogenes and is effective against only some races of a pathogen, whereas HR which is polygenic in inheritance is effective against all races of a pathogen. The term "major and minor genes" is sometime used instead of oligogenic and polygenic when referring to gene resistance. All oligogenes are not necessarily major genes in the sense of being important genes. For example, the R 1 gene governing resistance to potato blight merits being called major gene. But the gene R4 is weak and of little importance to potato breeders, and hence, it would be an exaggeration to call the R4 gene a major gene. In the same way, all polygenes are not minor genes in the sense of being relatively less important.

Methods of Selection of resistant Genotypes

While searching for resistant genotypes, a breeder has to look for the former by one of the following methods:

Selection from Existing Crops

This is done in the following ways:

Selection from Crops thac Escape Damage in Infected Fields The early breeders used this as one of the prominent methods, but it should be noted that such selected material cannot be usually called a genetically pure stock. As early as 1900, Orton of the US Department of Agriculture began selecting cotton varieties against cotton wilt *He* selected individual plants which showed resistance to disease in a wilt infested plot of land, that is even when they were drastically exposed to the disease. The seed from the selected plants was repeatedly' tested for resistance in progeny tests on the infected soil. By this process a number of good yielding wilt resistant varieties are involved

The wilt disease caused by *F. oxysporum* f. *sp.conglutinalls* has practically ruined cabbage production in parts of the USA.

The disease was investigated by L.R. Jones and others. Some seeds were collected from three highly resistant plants and resistant varieties of the Hollander type were developed. Further work at Wisconsin by L.R. Jones showed that temperature plays an important part in the onset of this disease. He found that some varieties became non-resistant at high temperature. He showed that 'when the soil was kept at 18.3°C or lower, plants grown in infected soil did not develop disease. As the temperature rose, the intensity of disease increased up to about 27°C after which it declined. Resistant varieties were selected on the lines given above. Later on, hybridization was also resorted to, and homozygous resistant varieties were evolved. This was the first experimental proof of the fact that in disease resistance, environment plays a very important part. Curly top of sugar beets, a virus disease which causes devastating losses has also been controlled by the production of resistant varieties by selection. The curly top resistant varieties were a spectacular success and saved the sugar beet crop in western USA. It should be emphasized here that the selection methods for resistant varieties can succeed only when the population has some resistant biotypes.

Pure Line Selection The above method is followed in pur~ line selection also but seed from selected plants is collected and tried separately. The stock thus produced is genetically pure.

Plant Introduction Very often, the genes needed to control a particular disease or race of a pathogen cannot be found in domestic varieties. It becomes necessary to introduce plant materials from other countries, especially from places where the pathogen and the host species may have evolved together and thus provide rich pools of resistance genes (Leppik, 1970).

Hybridization Among the existing varieties it is found that a variety may be resistant to a disease but may not possess suitable agricultural characteristics, that is, it may not be high yielding or the quality of the produce may not be satisfactory. On the other hand, there may be good varieties of high yields but they may be susceptible to disease. A breeder tries to combine the good characteristics of both the varieties to evolve a high yielding and resistant variety.

Hybridization:	H-High yield
	h-low yield
	R-Resistant
	r-Susceptible
HHrr	hhRR
(High yielding	(Low yielding
susceptible)	resistant
Gametes Hr×hR	

F1 generation=HhRr Phenotypically high yielding resistant

Gametes HR, Hr, hR, hr

	HR	Hr	hR	hr
HR	HHRR	HHRr	HhRR	HhRr
Hr	HHRr	HHrr	HhRr	Hhrr
hR	HhRR	Hhrr	hhRR	hhRr
hr	HhRr	Hhrr	hhRr	hhrr

(Out of these the first HHRR is high yielding and resistant and is a pure stock)

Selection from Wild Varieties

A search has often to be made for resistant genes from the wild varieties. The best places to find wild varieties resistant to certain pathogens are the primary and secondary, gene centres of cultivated plants. Vavilov used the term "gene centres" or centres of origin of primary regions of diverse cultivated species and varieties. He considered that these centre contain the greatest wealth of hereditary variants with definite naturally inbred character tics. Cultivated plants in these centres were frequently, but not always, accompanied closely related wild species as possible progenitors of domesticated plants. It soon became necessary to distinguish between primary and secondary gene centres. In the primary centre, the cultivated plants evolved from wild progenitors to distinguish between primary and secondary gene centres. In the primary centre, the cultivated plants evolved from wild progenitors native to these regions. During cultivation, however, these plants were frequently brought to other areas, where they were subject to different climatic and edaphic conditions. These climatic factors forced the plant population' to produce new diverse genetic characteristics. This is how secondary a tertiary centres of diversity were developed outside the primary centres.

Both the host and parasite have long been associated in their centres of origin reciprocal selective factors in evolution. Every new and more virulent race of para: necessarily eliminates most of the susceptible individuals in the local host population. 1 individuals that survive definitely carry genotypes different from these of the bulk of population. In this way, the survivors perpetuate genotypic resistance. Whenever serious diseases occur in nature, the highest degree of resistance develops among wild species: varieties as a result of natural selection.

It has, however, to be remembered that the gene centre of the host plant is also centre of origin of its specialized pests and pathogens, as well as the new and more virulent idiotypes, biotypes and pathogenic races which occur in the population of the parasites.

According to Vavilov there are nine world magacentres of origin of the D important cultivated plants in China, India (with a secondary Indo-Malayan center), Central Asia, Australia, the Near East, the Mediterranean region, North-east Africa Abyssinia, Southern Mexico and Central America, and South America (Peru-Equador Bolivia) with two subcentres (Chile and Brazil-Paraguay).

Although breeders have often turned to wild varieties of cultivated plants to find sources of resistance and have isolated stocks possessing the desired resistance, i difficult to transfer this resistance because of incompatibility problems between subject species or because of sterility in the hybrids. Breeders have now come to rely on special cytogenetic techniques and embryo culture to facilitate the more difficult crosses.

Induced Mutations New gene mutations can be introduced artificially with the help of abnormal radiations, such as x-rays, ultraviolet rays, etc. and also mutagenic chemicals like colchicine. ethyl methane sulphonate. The new mutations which arise may be useful or useless depending upon chance. The useful ones are selected and bred as per one of the above described methods.

Testing of Resistant Varieties

During the development of resistant varieties and before they are distributed to growers, it is necessary that they be tested under different environmental conditions for a sufficient amount of time to ensure durability of resistance. The following three precautions are important during testing:

a. The variety has been grown in selected areas of the region *or* country and the effect of various environments on its resistance has been determined to forecast when the variety will remain resistant and give the desired yield.

b. The physiological races of the pathogen prevalent in different regions have been taken into consideration.

c. The stability of resistance and other desired characteristics of the variety. have been established.

One of the main steps in testing resistant varieties is to create epiphytotic conditions. The aim is to ensure that significant inoculum comes in contact with the host under the, most favourable conditions for the pathogen. The various methods followed for some specific groups of diseases are briefly given as follows:

Soil-borne Diseases

Certain diseases are caused *by* soil-borne pathogens that enter the host plant through the roots or other underground parts. These include important and widespread diseases, such as wilt of cotton, *Rhizoctonia* rot of cotton, foot rot of wheat, wilt of pigeon pea, root rot of jute, flax wilt, etc. The varietal resistance to these ,diseases may be conducted *by* growing the varieties in soil in which the disease-inciting pathogen is prevalent. The intensity of the disease may be increased, in some cases by collecting soil *from* other diseased fields and scattering it over the test plot or *by* inoculating the soil with cultures of the causal organism grown in sand oatmeal medium or other nutrient media. The same soil area is then used again in succeeding years as a sick plot.

Seed-borne Diseases

Some smut diseases are externally seed-borne and inoculation techniques used for these diseases consist of applying the spores to the seed before planting the seeds as in loose, and covered smut of *Sorghum*.

Loose Smut of Wheat and Barley These smuts are the blossom infection type and infection in nature occurs at the time of flowering. The fungus mycelium gets established in the embryo. Seeds look apparently healthy but produce plants with smutted ears. .In such diseases the mature spores of the pathogen are brought in contact with flowers at the time of anthesis. This can be done with a fine forceps or hypodermic needle. If the contact is established under conditions of vacuum there are better chances of infection.

Rusts, Powdery Mildews and Blast of Rice Spores of rusts, powdery mildews and blast fungus are diseminated by wind, and under conditions of dim light, humidity and temperature cause infection through the stoma. Since some of these are obligate parasites, their inoculum cannot be multiplied on artificial media. Several procedures may be used *to* obtain better infections. Some of these are:

(a) Maintaining a temperature around the plants (host) during the infection' period that is optimum for the growth of the pathogen.

(b) Controlling humidity during the infection period by surrounding the *host* plant with a moist chamber in which the atmosphere can be maintained at or near saturation.

(c) Spraying the host plant with a suspension of the disease inciting organism during the days when the stomata are open widest so as to have the germinated spores or infection process in close proximity *to* the open stomata.

(d) Reducing the surface tension of the suspension to permit an even .spread of the inoculum by adding a small amount of a mild detergent or by subliming the waxy bloom off the leaves before dusting them with dry spores.

Insect-transmitted Diseases

Some diseases are transmitted by insects. Most of the virus diseases are transmitted by this means. Artificial inoculation techniques with these diseases may utilize the following procedures:

Insect Transmission. Insects, commonly aphids or leaf hoppers, which have fed on infected plants are collected and transferred to healthy plants. Plants are generally grown in insect proof cages *to* prevent natural movement of insects.

Mechanical Transfer. A diseased plant part is macerated and extracted juices rubbed over the leaves of healthy plants with sufficient force *to* cause slight mechanical injury. A fine abrasive, such as carborundum powder may be dusted over the leaves first, or mixed with juices, to aid in inflicting injury to the plant.